BEBOP TO THE BOOLEAN BOOGIE

An unconventional guide to electronics fundamentals, components, and processes

by Clive (call me "Max") Maxfield

Foreword by Pete Waddell,
Editor of Printed Circuit Design Magazine

HighText
publications inc.
Solana Beach, California

Library of Congress Cataloging-in-Publication Data

Maxfield, C. R. (Clive Richard), 1957–
    Bebop to the Boolean boogie : an unconventional guide to
electronics fundamentals, components, and processes / by C. R.
Maxfield : foreword by Pete Waddell.
        p.   cm
    Includes bibliographical references and index.
    ISBN: 1-878707-22-1 (paper) : $35.00
    1. Digital electronics—Popular works.   I. Title.
TK7868.D5M323   1995
821.39'5—dc20   94-41062
                                                                            CIP

Printed in the United States of America
Cover design: Brian McMurdo, Ventana Studio, Valley Center, CA
Interior design and production services: Sara Patton, Maui, HI

HighText is a registered trademark
of HighText Publications, Inc.

P.O. Box 1489 ✦ Solana Beach, CA, 92075

Foreword

My first exposure to the unique writing style of Clive (call me Max) Maxfield was a magazine article that he co-wrote with an associate. The article was technically brilliant (he paid me to say that) and very informative, but it was the short biography at the end of the piece that I enjoyed the most. I say enjoyed the most because, as you will soon learn, Max does not necessarily follow the herd or dance to the same drummer as the masses. Trade journals have a reputation for being informative and educational but also as dry as West Texas real estate.

Anyway, Max's personally submitted biography not only included a message from his mom, but also made mention of the fact that he (Max) is taller than his co-author, who just happened to be his boss at the time. Now to some people this may seem irrelevant, but to our readers (and Max's boss), these kind of things—trivial as they may seem to the uninitiated—are what helps us to maintain our off-grid sense of the world. Max has become, for better or worse, a part of that alternate life experience.

So now it's a couple of years later, and Max has asked me to write a few words by way of introduction. Personally, I think that the title of this tome alone (hmmm, a movie?) should provide some input as to what you can expect. But, for those who require a bit more: be forewarned, dear reader, you will probably learn far more than you could hope to expect from *Bebop to the Boolean Boogie*, just because of the unique approach Max has to technical material. The author will guide you from the basics through a mine field of potentially boring theoretical mish-mash, to a Nirvana of understanding. You will not suffer that fate familiar to every reader: rereading paragraphs over and over wondering what in the world the author was trying to say. For a limey, Max shoots amazingly well and from the hip, but in a way that will keep you interested and amused. If you are not vigilant, you may not only learn something, but you may even enjoy the process. The only further advice I can give is to "expect the unexpected."

– PETE WADDELL, Editor, *Printed Circuit Design*
Literary genius (so says his mom), and taller than Max by ⅛"

Contents

Figures

Acknowledgments

Had I known how difficult it was going to be to write this book, I would probably have taken up down-hill mud wrestling[1] instead. The only reason that I eventually succeeded is due to the tremendous support that I received from my family and friends.

Special thanks for technical advice are due to Alvin Brown, Alon Kfir, Don Kuk, and Preston Jett,[2] the closest thing to living encyclopedic reference manuals one could hope to meet. The reason that the text contains so few bibliographic references is due to the fact that I never had to look anything up—I simply strolled around the corner and asked the relevant expert for the definitive answer.[3] Also, thanks go to Dave Allen, Pat Williams, Rob Kunkle, and Ed Smith for their sterling advice on programmable logic, seafood gumbo, circuit boards, and C programming, respectively.

Thanks are also due to my wife Stephanie for the many hours she spent laboring over the formatting, and to my small army of reviewers: Tony McCollum, Dana Stephenson, Dan Polfer, Sam Lay, and James Johnson. Additionally, I am very grateful for the painstaking work put forward by Barbara Castlen and my mother, Margaret Maxfield, in the final grammatical check. In the unlikely event that any errors did creep in, they can only be attributed to cosmic rays and have nothing to do with me.

I should also like to mention my daughters, Abby and Lucie, without whom this book would never have materialized. They so depleted my financial resources that I was obliged to look for a supplemental source of income.

---

[1] Always a personal favorite.

[2] To avoid any hint of preferential treatment, the cast appears in reverse alphabetical sort of the third character in the first name.

[3] This was particularly useful because, although electronic concepts and practices are defined and documented almost to the point of absurdity, many of them tend to be defined and documented in various, and often contradictory, ways.

Finally, who could avoid acknowledging my friend (I use the term in the loosest sense) Al Gilbert. His "last of the reverse-mohican haircuts" and technicolor suspenders[4] are a source of endless amusement, even on the darkest days.

– CLIVE (MAX) MAXFIELD
July 1992 – April 1994

---

[4] On the off-chance that Al is ever allowed out in public, it should perhaps be explained that suspenders in America are used to hold one's trousers up, as opposed to an article of feminine lingerie, which would be the case in England.

About the Author

Clive (Max) Maxfield lives with three women. However, as one of them is his wife, Stephanie, and the others are his daughters, Lucie and Abby, this is not quite as exciting as it might at first appear. Max is 6'1" tall, outrageously handsome, English and proud of it. In addition to being a trend setter and a leader of fashion, he is widely regarded as an expert in all aspects of electronic engineering (at least by his mother). Apropos of nothing at all, Max has a younger brother, Andrew (Max) Maxfield,[1] who would be tickled pink to see his name in a real book.

For those who are interested in this sort of thing, Max received a B.Sc. in Control Engineering from Sheffield Polytechnic, England, in 1980. He began his career as a mainframe CPU designer for International Computers Limited (ICL) in Manchester, England. To cut a long story short, Max now finds himself a member of the technical staff (MTS) at Intergraph Electronics, Huntsville, Alabama, USA. Between then and now, he has wandered through integrated circuit and circuit board design, sauntered through in-circuit and functional test programming, and hurled himself into design capture and simulation technologies.

Max is the author of dozens of articles and papers appearing in magazines and at technical conferences around the world. He was once referred to as an *"industry notable"* and a *"semiconductor design expert"* by someone famous who wasn't prompted, coerced, or renumerated in any way! Max's main areas of interest are currently focused in the analog, digital, and mixed-signal simulation of integrated circuits and multichip modules, but who knows what tomorrow will bring?

---

[1] Everyone in Max's family shares the same nickname which can result in somewhat convoluted after dinner conversations.

Dedication

This book is dedicated to my Auntie Barbara,
whose assiduous scrubbing in my younger years has
left me the proud owner of the cleanest pair of
knees in the known universe!

Introduction

This outrageously interesting book has two namesakes: Bebop, a jazz style known for its fast tempos and agitated rhythms, and Boolean algebra, a branch of mathematics that is the mainstay of the electronics designer's tool chest. *Bebop to the Boolean Boogie* meets the expectations set by both, because it leaps from topic to topic with the agility of a mountain goat, and it will become your key reference guide to understanding the weird and wonderful world of electronics. As Captain James T. Kirk would say: "*Bebop to the Boolean Boogie* boldly goes behind the beyond, behind which few books have boldly gone beyond before."

There are many excellent books on electronics, ranging from introductory texts to highly detailed reference works. However, most of these offerings focus on specific subjects, making it difficult to see the 'big picture.' By contrast, *Bebop to the Boolean Boogie* provides a wide-ranging but comprehensive introduction to the electronics arena, roaming through the fundamental concepts and rampaging through electronic components and the processes used to create them. As a bonus, nuggets of trivia are included with which you can amaze your family and friends; for example, Greenland Eskimos have a base twenty number system because they count using both fingers and toes.

When reading this book you should ideally start at the beginning, wander through the middle, and eventually arrive somewhere in the vicinity of the end. But each chapter has been written to be as self-sufficient as possible, and the style of the book is intended to facilitate your plunging in and out as you wish.

In some respects, the order and fashion in which information is presented differs from the format associated with that of traditional text books. However, the presentation format has been carefully selected, the chapters are short and focused, and their ordering provides an ongoing mixture of subjects. The book consists of two main sections which, taken together, span an exceptionally wide range:

Section 1: Fundamental Concepts

After considering the differences between analog and digital views of the world, the discussion progresses rapidly through atomic theory and semiconductor switches to primitive logic functions and their electronic implementations. The concepts of alternative numbering systems are presented, along with binary arithmetic, Boolean algebra, and Karnaugh map representations. Finally, the construction of more complex logical functions is considered along with their applications.

Section 2: Components and Processes

This section considers the components from which electronic systems are formed and the processes required to construct them. The construction of integrated circuits is examined in some detail, followed by introductions to memory devices, programmable devices, and application-specific devices. The discussion continues with hybrids, printed circuit boards, and multichip modules, and closes with an overview of some alternative and future technologies.

This book is of particular interest to first-year electronics students. Additionally, by clarifying the techno-speech used by engineers, the book is of value to anyone who is interested in understanding more about electronics but lacks a strong technical background.

Except where such interpretation is inconsistent with the context, the singular shall be deemed to include the plural, the masculine shall be deemed to include the feminine, and the spelling shall be deemed to be correct.

A Brief History of Computers and Electronics

Before moving on to the "meat of the book," it is appropriate to first consider a brief history of computers and electronics to set the stage for all that is to follow.

No one person may be credited with the invention of electronics or computers, but several names stand proud in the crowd. The following attempts to cover at least a few of the more notable individual developments, with great regret for any omissions.

The first tools used to aid calculation were almost certainly man's own fingers, and it is not simply a coincidence that the word *"digit"* is used to refer to a finger (or toe) as well as a numerical quantity. As the need to represent larger numbers grew, early man employed readily available materials for the purpose. The oldest known objects used to represent numbers are bones with notches discovered in western Europe. Of special interest is a wolf bone more than 20,000 years old with 55 notches in groups of five, which was discovered at Vestonice, Czechoslovakia in 1937. Also of use were small stones or pebbles, which could represent larger numbers than fingers and toes and have the advantage of being able to easily store intermediate results for later use. Thus, it is also no coincidence that the word *"calculate"* is derived from the Latin word for pebble.

The first actual calculating mechanism known to us[1] is the abacus, which is thought to have been invented sometime between 1,000 BC and 500 BC. There are a number of versions of abacus, but the most popular are based on the bi-quinary system, which utilizes two bases, base-2 and base-5, to represent decimal numbers.

The great Italian Renaissance man *Leonardo da Vinci* (1452–1519) was a genius: painter, musician, sculptor, architect, engineer, etc. However, his contributions to mechanical calculation remained hidden until the rediscovery of two of his notebooks in 1967. Working models of da Vinci's calculator have since been constructed from the drawings in these notebooks.

In the early 1600s, a Scottish mathematician, *John Napier* (1550–1617), invented a tool called Napier's Bones, which were multiplication tables written out on strips of wood or bone. Napier also invented logarithms, which greatly assisted in arithmetic calculations. The combination of logarithms and Napier's bones formed the basis for the modern slide rule.

A French mathematician, physicist, and theologian, *Blaise Pascal* (1623–1662), is credited with the invention of the first operational calculating machine. In 1640 Pascal started developing a device to help his father

---

[1] It is more than possible that other similar inventions have been lost in the mists of time.

add sums of money in his tax work. The first operating model, the *Arithmetic Machine*, was introduced in 1642 and Pascal created fifty more devices over the next ten years.[2]

Gottfried von Leibniz (1646–1716) entered university at 15 years of age and received his bachelor's degree at 17. Leibniz said: "It is unworthy of excellent men to lose hours like slaves in the labor of calculation which could be safely relegated to anyone else if machines were used." Leibniz developed Pascal's ideas and, in 1671, introduced the *Step Reckoner*, a device which could multiply, divide, and evaluate square roots by series of stepped additions. Pascal's and Leibnitz's devices were the forebears of today's desktop computers.

In the early 1880s, a French silk weaver called *Joseph Jacquard* (1752–1834) invented a way of automatically controlling the warp and weft threads on a silk loom by recording patterns as holes in a string of cards. In the years to come, variations on Jacquard's punched cards found a variety of uses, including the storage of music for automated pianos and the storage of programs for computers.

The first device that might be considered to be a computer in the modern sense was conceived by the eccentric British mathematician *Charles Babbage* (1791–1871). In Babbage's time, mathematical tables such as logarithmic and trigonometric functions were generated by teams of mathematicians working night and day on primitive calculators. In 1822, Babbage proposed building a machine called the *Difference Engine* to automatically calculate these tables.

The Difference Engine was partially completed when Babbage had the idea for another machine that would use Jacquard's punched cards. The proposed *Analytical Engine* was to use loops of cards to control an automatic calculator and make decisions based on the results of previous computations. The Analytical Engine was intended to employ several features subsequently used in modern computers: sequential control, branching, and looping.

---

[2] In 1658, Pascal created a scandal when, under the pseudonym of Amos Dettonville, he challenged other mathematicians to a contest and then awarded the prize to himself!

Working with Babbage was the daughter of Lord Byron, *Augusta Ada Lovelace* (1815–1852), who created a program for the Analytical Engine that would compute Bernoulli numbers.[3] Based on this work, Ada is now credited as being the first computer programmer and, in 1979, the modern programming language ADA was named in her honor.

Babbage worked on the Analytical Engine from around 1830 until he died. Unfortunately, he was a hundred years ahead of his time; the technology of the day was inadequate for the task and the machine was never completed. Interestingly enough, one of Babbage's earlier Difference Engines was finally constructed by a team at London's Science Museum working from original drawings. The final machine consists of 4,000 components constructed from cast iron, bronze and steel, weighs three tons, and is 3.3 meters wide by 0.45 meters deep by 2.1 meters tall. It performed its first sequence of complex calculations in the early 1990s, returning results to 31 digits of accuracy, which is far more accurate than the standard pocket calculator. However, each calculation requires the operator to turn a crank hundreds, sometimes thousands of times.

Not too far away from Babbage, another British mathematician, *George Boole* (1815–1864), was busily inventing a new form of mathematics. Boole made significant contributions in several areas of mathematics, but was immortalized for two works in 1847 and 1854 in which he represented logical expressions in a mathematical form now known as *Boolean Algebra*. Boole's work was all the more impressive because, with the exception of elementary school and a short time in a commercial school, he was almost completely self-educated.

In conjunction with Boole, yet another British mathematician, *Augustus DeMorgan* (1806–1871), formalized a set of logical operations now known as *DeMorgan transformations*.[4] As the *Encyclopedia Britannica* says: "A renascence of logical studies came about almost entirely because

---

[3] In their spare time, Babbage and Ada also attempted to create a system for predicting the winners of horse races, but it is said that they lost a lot of money!

[4] In fact, the rules we ascribe to DeMorgan were known in a more primitive form by William Ockham in the 14th century.

of Boole and DeMorgan." Unfortunately, with the exception of students of symbolic logic, Boolean Algebra was destined to remain largely unknown and unused for the better part of a century.

The pace began to pick up somewhat in the twentieth century. In 1906, the American inventor *Lee de Forest* (1873–1961) introduced a third electrode called the grid into the thermionic valve. The resulting *triode* could be used as both an amplifier and a switch.

In 1921, the Czech author *Karel Capek* (1890–1938) produced his best-known work, the play R.U.R. (Rossum's Universal Robots), which featured machines created to simulate human beings. The term robot came from the Czech word *robota*, meaning work.

In the late 1930s, a graduate student at MIT, *Claude E. Shannon*, realized that Boolean Algebra provided an ideal mathematical system for designing computers. His Master's thesis showed how Boole's concepts of TRUE and FALSE could be used to represent the functions of switches in electronic circuits. Shannon's paper has been described as: "Possibly the most important Master's thesis of the century." Shannon is also credited with the invention of the rocket-powered frisbee and is famous for riding down the corridors at Bell Laboratories on a unicycle while simultaneously juggling four balls.

Many consider the modern computer era to commence with the first large-scale automatic digital computer developed between 1939 and 1944; the *Harvard Mark I*. The Mark I was primarily constructed out of switches and relays and was described as sounding like a "roomful of ladies knitting." The machine was 51 feet long, 8 feet high, and weighed approximately 5 tons! Although the Mark I is considered to be the first digital computer, its architecture was significantly different from modern machines. The device consisted of many calculators working on parts of the same problem under the guidance of a single control unit. Instructions were read in on paper tape, and the device could only perform operations in the sequence in which they were received.

In the aftermath of World War II, it was discovered that a program-controlled calculator called the Z3 had become functional in Germany in

1941. Developed by *Konrad Zuse*, the Z3 therefore predates the Mark I. Some believe that the Z3 was faster than the Mark I, but unfortunately it didn't survive the war.

By the mid-1940s, computers were being built out of vacuum tubes rather than switches and relays. Although the vacuum tubes were fragile, expensive, and used a lot of power, they were much faster than relays and much quieter. The first true electronic computer, *ENIAC* (Electronic Numerical Integrator And Calculator), was constructed at the University of Pennsylvania between 1943 and 1946. In many ways ENIAC was a monster; it occupied 30 feet by 50 feet of floor space, weighed approximately 30 tons, and used more than 18,000 vacuum tubes requiring 150 kilowatts of power, enough to light a small town. One of the greatest problems with computers built from vacuum tubes was reliability; 90% of ENIAC's down-time was attributed to locating and replacing burnt-out tubes. Records from 1952 show that approximately 19,000 vacuum tubes had to be replaced in that year alone; this averages out to about 50 tubes a day!

One of the consultants to the ENIAC development group was the Hungarian-American mathematician *Johaan Von Neumann* (1903–1957). Although no one person may be credited with inventing the computer, many consider Von Neumann to be the single most important contributor. During the development of ENIAC, Von Neumann conceived the idea of storing instructions in memory along with the data. Rather than simply following a predetermined sequence of instructions, this would allow the program to branch to alternate instruction sequences based on the results of previous calculations. Additionally, as the electronic vacuum tubes were so much faster than relays, it was no longer necessary to use multiple computing elements operating in parallel.

Von Neumann gathered his ideas together in a paper issued in June 1945, in which he presented the basic elements of a *stored-program* computer:

    i.   to have a memory containing both data and instructions, and to allow both data and instruction memory locations to be read from, and written to, in any desired order;

ii.   to have a calculating unit capable of performing both arithmetic and logical operations on the data;

iii.   and to have a control unit which could interpret an instruction received from the memory and select alternative courses of action based on the results of previous operations.

The key point was that the computer could modify its own programs in much the same way as was originally suggested by Babbage. The computer structure resulting from Von Neumann's criteria is popularly known as a Von Neumann Machine, and virtually all digital computers from that time have been based on this architecture.

Apropos of nothing at all, a jazz style known as *Bebop* became highly popular in the decade following World War II. *Charlie Parker, Dizzy Gillespie* and *Thelonius Monk* were especially associated with this form of music, known for its fast tempos and agitated rhythms. We may only speculate if it was but a coincidence that many of the most significant ideas and discoveries in the history of computing occurred alongside the flourishing Bebop.

Von Neumann's ideas led to the proposal for the first stored program machine, *EDVAC* (Electronic Discrete Variable Automatic Computer), which was developed at the University of Pennsylvania. Unfortunately, although the conceptual design was completed by 1946, several key members left the project to pursue their own careers and the machine did not become fully operational until 1952. EDVAC contained approximately 4,000 vacuum tubes and 10,000 crystal diodes. A 1956 report shows that EDVAC's average error-free up-time was approximately 8 hours.

In fact, some would dispute EDVAC's claim-to-fame as the first stored program computer due to its late completion. Based on the ideas set forward for EDVAC, a small experimental machine comprising 32 words of memory and a 5-instruction instruction-set was operating at Manchester University, England, by June 1948. Additionally, a machine called *EDSAC* (Electronic Delay Storage Automatic Calculator) performed its first calculation at Cambridge University, England, in May 1949.

The first commercially available computer, *UNIVAC I* (Universal Automatic Computer), was also based on the EDVAC design. Work started on UNIVAC I in 1948, and the first unit was delivered in 1951. Once again, due to the late completion of EDVAC, the first UNIVAC unit was delivered a year before EDVAC was fully operational.

The transistor and subsequently the integrated circuit must certainly qualify as two of the greatest inventions of the twentieth century. Driven by research on radar in World War II, scientists and engineers began to investigate a little known group of substances called semiconductors. It was considered that devices formed from semiconductors had potential as amplifiers and switches and could therefore be used to replace the prevailing technology of vacuum tubes.

Bell Laboratories in the United States began research into semiconductors in 1945, and physicists *William Shockley*, *Walter Brattain* and *John Bardeen* succeeded in creating the first point-contact germanium transistor on December 23, 1947. In 1950, Shockley invented a new device called a bipolar junction transistor that was more reliable, easier and cheaper to build, and gave more consistent results than point-contact devices.

By the late 1950s, bipolar transistors were being manufactured out of silicon rather than germanium; in addition to being easier to work with, silicon offered certain electrical advantages. Bipolar junction transistors are formed from the junction of three pieces of doped silicon called the collector, base, and emitter. The original bipolar transistors were manufactured using the mesa process, in which a doped piece of silicon called the mesa (or base) was mounted on top of a larger piece of silicon which formed the collector; the emitter was created from a smaller piece of silicon embedded in the base.

In 1959, the Swiss Physicist *Jean Hoerni* invented the planar process in which optical lithographic techniques were used to diffuse the base into the collector, then the emitter into the base. One of Hoerni's colleagues, *Robert Noyce*, invented a technique for growing a layer of silicon dioxide insulator over the transistor, leaving small areas over the base and emitter exposed, and diffusing thin layers of aluminum into these areas to create wires. The processes developed by Hoerni and Noyce led directly to modern integrated circuits.

In 1962, *Steven Hofstein* and *Fredric Heiman* at the RCA research laboratory in Princeton, New Jersey, invented a new family of devices called metal-oxide semiconductor field-effect transistors. Although these transistors were somewhat slower than bipolar transistors, they were cheaper, smaller and used less power. Additionally, modified metal-oxide semiconductor structures could be made to act as capacitors or resistors.

To a large extent the demand for miniaturization was driven by the demands of the American space program. For some time people had been thinking that it would be a good idea to be able to fabricate entire circuits on a single piece of semiconductor. The first public discussion of this idea is credited to a British radar expert, *G.W.A. Drummer*, in a paper presented in 1952. However, it was not until the summer of 1958 that Jack Kilby, working for Texas Instruments, succeeded in fabricating multiple components on a single piece of semiconductor. Kilby's first prototype was a phase shift oscillator and, although manufacturing techniques subsequently took different paths from those used by Kilby, he is still credited with the manufacture of the first true integrated circuit.

By 1961, Fairchild and Texas Instruments had announced the availability of the first commercial planar integrated circuits comprising simple logic functions. This announcement marked the beginning of the mass production of integrated circuits. In 1963, Fairchild produced a device called the 907 containing two logic gates, each of which consisted of four bipolar transistors and four resistors. The 907 also made use of isolation layers and buried layers, which were to become common features in modern integrated circuits.

In 1967, Fairchild introduced a device called the *Micromosaic* which contained a few hundred transistors. The key feature of the Micromosaic was that the transistors were not initially connected to each other. A designer used a computer program to specify the function the device was required to perform, and the program determined the necessary transistor interconnections and constructed the masks required to complete the device. The Micromosaic is credited as the forerunner of the modern application-specific integrated circuit and the first real application of computer-aided design.

In 1970, Fairchild introduced the first 256-bit static RAM called the 4100; in the same year Intel announced the first 1024-bit dynamic RAM called the 1103. Then, in 1971, Intel presented the first *microprocessor* to reach the market, the 4004. This 4-bit device was part of a four-chip system which also comprised a 256-byte ROM, a 32-bit RAM, and a 10-bit shift register. The 4004 contained approximately 2,300 transistors and could perform 60,000 operations per second. The 4004 was followed in 1972 by the 8008, an 8-bit microprocessor which contained approximately 3,300 transistors. However, both the 4004 and 8008 microprocessors were designed for specific applications. In 1974, Intel presented the first true general-purpose microprocessor, the 8-bit 8080, which contained about 4,500 transistors and could perform 200,000 operations per second.

The advent of the general-purpose microprocessor heralded a new era in computing. Systems small enough to fit on a desk could be created with more processing power than monsters weighing tens of tons only a decade or two before. Almost unbelievably, individuals could now own a personal computer. The effects of these developments are still unfolding, but it is not excessive to say that the digital computer has changed the world more significantly than almost any other human invention.

Analog versus Digital

Now sit up and pay attention because this bit is important. Electronic engineers split their world into two views, *analog* and *digital*, and it's necessary to understand the difference between these views to make much sense out of the rest of this book. Consider a fun-loving fool sliding down a ramp mounted alongside a staircase (Figure 1.1).

Figure 1.1: Staircase and ramp

To accurately determine the person's position on the ramp, an independent observer would require the use of a tape measure. Alternatively, the observer could estimate the ramp-slider's approximate location in relation to the nearest stair. The exact position on the ramp, as measured using the tape measure, would be considered to be an analog value. An analog value most closely represents the real world and can be as precise as the measuring technique allows. By comparison, an estimation based on the nearest stair would be considered to be a digital value. A digital value is represented as being in one of a finite number of discrete states called *quanta*. The accuracy, or resolution, of a digital value is dependent on the number of quanta employed to represent it.

Assume that at some starting time called T_O our thrill-seeker is balanced at the top of the ramp preparing to take the plunge. He commences sliding at time T_1 and reaches the bottom at time T_2. Graphs can be plotted representing the location of the person on the ramp as a function of time; these representations are known to electronic engineers as *waveforms* (Figure 1.2).

Figure 1.2: Person on a ramp: analog and digital waveforms

The horizontal axis in both waveforms represents the passage of time, which is considered to progress from left to right. In the case of the analog waveform, the vertical axis is used to represent the thrill-seeker's exact location in terms of height above the ground, and is therefore annotated with real, physical units. By comparison, the vertical axis for the digital waveform is annotated with abstract labels which do not have any units associated with them.

To examine the differences between the analog and digital views in more detail, consider a brick suspended from a wooden beam by a piece of elastic.

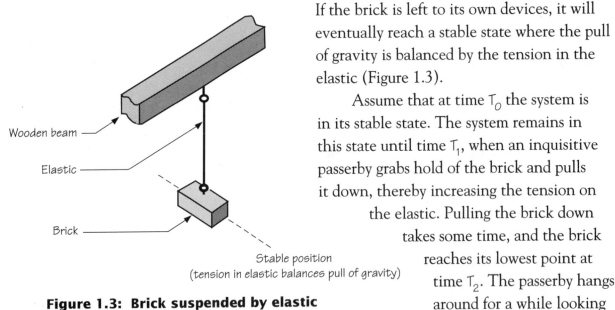

If the brick is left to its own devices, it will eventually reach a stable state where the pull of gravity is balanced by the tension in the elastic (Figure 1.3).

Assume that at time T_0 the system is in its stable state. The system remains in this state until time T_1, when an inquisitive passerby grabs hold of the brick and pulls it down, thereby increasing the tension on the elastic. Pulling the brick down takes some time, and the brick reaches its lowest point at time T_2. The passerby hangs around for a while looking

Figure 1.3: Brick suspended by elastic

somewhat foolish, releases the brick at time T_3, and thereafter exits from our story. The brick's resulting motion may be illustrated using an analog waveform:

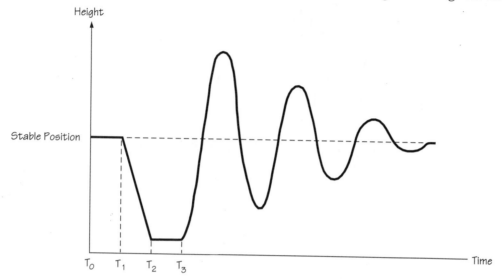

Figure 1.4: Brick on elastic: analog waveform

Now consider a digital view of the brick's motion represented by two quanta labeled LOW and HIGH. The LOW quanta may be taken to represent any height less than or equal to the system's stable position, and the HIGH quanta therefore represents any height greater than the stable position:

Figure 1.5: Brick on elastic: digital waveform (two quanta)

The analog waveform is only shown here for reasons of clarity. Although it is apparent that the digital view is a subset of the analog view, digital representations often provide extremely useful approximations to the real world. If the only requirement in the above example is to determine whether the brick is above or below the stable position, then the digital view is the most appropriate.

The accuracy of a digital view can be improved by adding more quanta. For example, consider a digital view with five quanta: LOW, LOW-MIDDLE, MIDDLE, HIGH-MIDDLE, and HIGH. As the number of quanta increases, the digital view more closely approximates the analog view:

Figure 1.6: Brick on elastic: digital waveform (five quanta)

In the real world, every electronic component behaves in an analog fashion. However, these components can be connected together so as to form functions whose behavior is amenable to digital approximations. This book concentrates on the digital view of electronics, although certain aspects of analog designs and the effects associated with them are discussed where appropriate.

Atoms, Molecules, and Crystals

Matter, the stuff that everything is made of, is formed from *atoms*. The heart of an atom, the *nucleus*, is composed of *protons* and *neutrons* and is surrounded by a "cloud" of *electrons*. It may help to visualize electrons as orbiting the nucleus in the same way that the moon orbits the earth. In the real world things aren't quite this simple, but the concept of orbiting electrons serves our purpose:

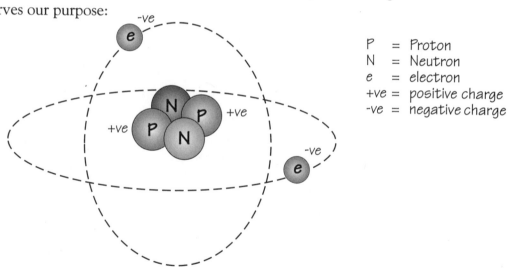

P = Proton
N = Neutron
e = electron
+ve = positive charge
-ve = negative charge

Figure 2.1: Helium atom

Each proton carries a single positive (+*ve*) charge and each electron carries a single negative (−*ve*) charge. The neutrons are neutral and act like glue, holding the nucleus together and resisting the natural tendency of the protons to repel each other. Protons and neutrons are approximately the same size, while electrons are very much smaller. If a basketball were used to represent the nucleus of a helium atom then, on the same scale, softballs could represent the individual protons and neutrons, while large garden peas could represent the electrons. In this case, the diameter of the electron's orbit would be approximately equal to the length of 250 football fields! (That's American football fields excluding the end zones.) Thus, the majority of an atom consists of empty space. If all the empty space were removed from the atoms that form a camel, it *would* be possible for a camel to pass through the eye of a needle![1,2]

The number of protons determines the type of the element; for example, hydrogen has one proton, helium two, lithium three, etc. Atoms vary greatly in size, from hydrogen with its single proton to those containing hundreds of protons. The number of neutrons does not necessarily equal the number of protons. There may be several different flavors, or *isotopes*, of the same element differing only in their number of neutrons; for example, hydrogen has three isotopes with zero, one, and two neutrons, respectively.

Left to their own devices, each proton in the nucleus will have a complementary electron. If additional electrons are forcibly added to an atom, the result is a *negative ion* of that atom; if electrons are forcibly removed from an atom, the result is a *positive ion*.

In an atom where each proton is balanced by a complementary electron, one would assume that the atom would be stable and content with its own company; however, things are not always as they seem. Although every electron contains the same amount of negative charge, electrons orbit the nucleus at different levels known as *quantum levels* or *electron shells*. Each electron shell requires a specific number of electrons to fill it; the first shell requires two electrons, the second requires eight, etc. Thus, although a hydrogen atom contains both a proton and an electron and is therefore electrically balanced, it is still not completely happy. Given a choice, hydrogen would prefer to have a second electron to fill its first electron shell. However, simply adding a second electron is not the solution; although the first electron shell would now be filled, the extra electron would result in an electrically unbalanced negative ion.

---

[1] I am of course referring to the Bible verse: "It is easier for a camel to go through the eye of a needle, than for a rich man to enter into the Kingdom of God" (Mark 10:25).

[2] This is irrespective of the size, age, weight, or even the name of the camel!

Obviously this is a bit of a stumper, but the maker of the universe came up with a solution; atoms can use the electrons in their outermost shell to form bonds with other atoms. The atoms share each other's electrons, thereby forming more complex structures. One such structure is called a *molecule*; for example, two hydrogen atoms (chemical symbol H), each comprising a single proton and electron, can bond together and share their electrons to form a hydrogen molecule (chemical symbol H_2):

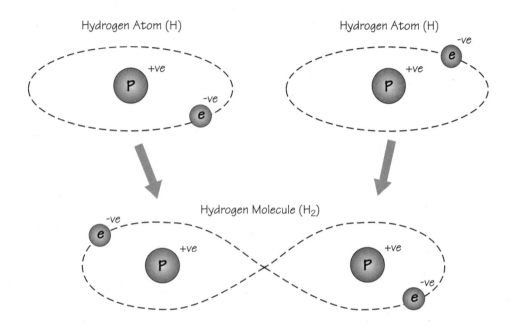

Figure 2.2: Two hydrogen atoms bonding to form a hydrogen molecule

These types of bonds are called *valence bonds*. The resulting hydrogen molecule contains two protons and two electrons from its constituent atoms and so remains electrically balanced. However, each atom lends its electron to its partner and, at the same time, borrows an electron from its partner. This can be compared to two circus jugglers passing objects between each other— the quickness of the hand deceives the eye. The electrons are passing backwards and forwards between the atoms so quickly that each atom is fooled into believing it has two electrons. The first electron shell of each atom appears to be completely filled and the hydrogen molecule is therefore stable.

Even though the hydrogen molecule is the simplest molecule of all, the previous illustration demanded a significant amount of time, space, and effort. Molecules formed from atoms containing more than a few protons and electrons would be well nigh impossible to represent in this manner. A simpler form of representation is therefore employed, with two dashed lines indicating the sharing of two electrons:

H_2

H)==== (H   H = hydrogen atom

Figure 2.3: Alternative representation of a hydrogen molecule

Now contrast the case of hydrogen with helium. Helium atoms each have two protons and two electrons and are therefore electrically balanced. Additionally, as helium's two electrons completely fill its first electron shell, this atom is very stable.[3] Under normal circumstances helium atoms do not go around casually making molecules with every other atom they meet.

Molecules can also be formed by combining different types of atoms. An oxygen atom (chemical symbol O) contains eight protons and eight electrons. Two of the electrons are used to fill the first electron shell, which leaves six left over for the second shell. Unfortunately for oxygen, its second shell would ideally prefer eight electrons to fill it. Each oxygen atom can therefore form two bonds with other atoms—for example, with two hydrogen atoms to form a water molecule (chemical symbol H_2O) (Figure 2.4). The reason

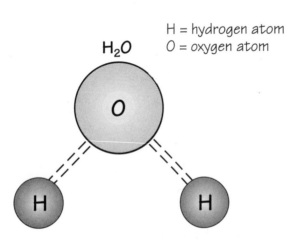

H_2O

H = hydrogen atom
O = oxygen atom

Figure 2.4: Water molecule

---

[3] Because helium is so stable, it is known as an *inert*, or *noble*, gas.

the three atoms in the water molecule are not shown as forming a straight line is discussed in the section on nanotechnology in Chapter 21.

Each hydrogen atom lends its electron to the oxygen atom and at the same time borrows an electron from the oxygen atom. This leads both of the hydrogen atoms to believe they have two electrons in their first electron shell. Similarly, the oxygen atom lends two electrons (one to each hydrogen atom) and at the same time borrows two electrons (one from each hydrogen atom). When the two borrowed electrons are added to the original six in the oxygen atom's second shell, this shell appears to contain the eight electrons necessary to fill it. Thus, all the atoms in the water molecule are satisfied with their lot and the molecule is stable.

Structures other than molecules may be formed when atoms bond; for example, *crystals*. Carbon, silicon and germanium all belong to the same family of elements; each has only four electrons in its outermost electron shell. Silicon has 14 protons and 14 electrons; two electrons are required to fill the first electron shell and eight to fill the second shell; thus, only four remain for the third shell, which would ideally prefer eight. Under the appropriate conditions, each silicon atom will form bonds with four other silicon atoms, resulting in a three-dimensional silicon crystal[4] (Figure 2.5).

The electrons used to form the bonds in crystalline structures such as silicon are tightly bound to their respective atoms. Yet another structure is presented by metals such as copper, silver, and gold. Metals have an amorphous crystalline

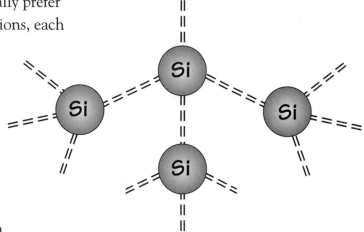

Si = silicon atom

Figure 2.5: Three-dimensional structure of silicon crystal

---

[4] An equivalent structure formed from carbon atoms is known as diamond.

structure in which their shared electrons have relatively weak bonds and may easily migrate from one atom to another.

Apart from the fact that atoms are the basis of life, the universe, and everything as we know it, they are also fundamental to the operation of the components used in electronic designs. Electricity may be considered to be vast herds of electrons migrating from one place to another, while electronics is the science of controlling these herds: starting them, stopping them, deciding where they can roam, and determining what they are going to do when they get there.

Conductors and Insulators; Voltage, Current, Resistance, Capacitance, and Inductance

A substance which conducts electricity easily is called a *conductor*. Metals such as copper are very good conductors because the bonds in their amorphous crystalline structures are relatively weak, and the bonding electrons can easily migrate from one atom to another. If a piece of copper wire is used to connect a source with an excess of electrons to a target with too few electrons, the wire will conduct electrons between them:

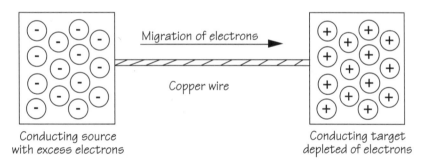

Figure 3.1: Electrons flowing through a copper wire

If we consider electricity to be the migration of electrons from one place to another, then we may also say that it flows from the more negative source to the more positive target. As an electron *jumps* from the negative source into the wire, it *pushes* the nearest electron in the wire out of the way. This electron pushes another in turn, and the effect ripples down the wire until an electron at the far end of the wire is ejected into the more positive target. When an electron arrives in the positive target it neutralizes one of the positive charges.

An individual electron will take a surprisingly long time to migrate from one end of the wire to the other; however, the time between an electron entering one end of the wire causing an equivalent electron to be ejected from the other end is extremely fast.[1]

As opposed to a conductor, a substance which does not conduct electricity easily is called an *insulator*. Materials such as rubber are very good insulators because the electrons used to form bonds are tightly bound to their respective atoms.

Voltage, Current and Resistance

One measure of whether a substance is a conductor or an insulator is how much it resists the flow of electricity. Imagine a tank of water to which two pipes are connected at different heights; the water ejected from the pipes is caught in two buckets A and B (Figure 3.2).

Assume that the contents of the tank are magically maintained at a constant level. The water pressure at the end of a pipe inside the tank depends on the depth of the pipe with respect to the surface level. The difference in pressure between the ends of a pipe inside and outside the tank causes water to flow. The amount of water flowing through a pipe depends on the water pressure and on the resistance to that

Figure 3.2: Water tank representation of voltage, current, and resistance

---

[1] For a copper wire in a vacuum, the speed of a signal propagating through the wire is only fractionally less than the speed of light. However, the speed of a signal is modified by a variety of factors, including any materials surrounding or near the conductor. The speed of signals in electronic circuits vary, but are typically in the vicinity of half the speed of light.

pressure determined by the pipe's cross-sectional area. A thin pipe with a smaller cross-sectional area will present more resistance to the water than a thicker pipe with a larger cross-sectional area. Thus, if both pipes have the same cross-sectional area, bucket B will fill faster than bucket A.

In electronic systems the flow of electricity is called *current* measured in units of amperes or amps,[2] the resistance to electrical flow is simply called *resistance* measured in units of ohms, and the electrical equivalent to pressure is called *voltage*, or electric potential, measured in units of volts.

The materials used to connect components in electronic circuits are typically selected to have low resistance values; however, in some cases engineers make use of special resistive components called *resistors*. The value of resistance (R) depends on the resistor's length, cross-sectional area, and the material from which it is formed. Resistors come in many shapes and sizes; an example could be as shown in Figure 3.3.

In a steady-state system where everything is constant, the voltage, current, and resistance are related by a rule called *Ohm's Law*, which states that voltage (V) equals current (I) multiplied by resistance (R). A simple method for remembering Ohm's Law is by means of a diagram known as Ohm's triangle:

Approx. actual size

(a) Discrete Component

(b) Symbol

Figure 3.3: Resistor: component and symbol

Figure 3.4: Ohm's triangle

---

[2] An amp corresponds to approximately 6,250,000,000,000,000,000 electrons per second flowing past a given point in an electrical circuit. Not that the author counted them himself, you understand; this little nugget of information is courtesy of Microsoft's multimedia encyclopedia, *Encarta*.

Ohm's Law: V = I × R
           I = V/R
           I = 5 volts/10 ohms
           I = 0.5 amps

Figure 3.5: Current flowing through a resistor

Consider a simple electrical circuit comprising two wires with electrical potentials of 5 volts and 0 volts connected by a resistor of 10 ohms (Figure 3.5). This illustration shows the direction of current flow as being from the more positive (+5 volts) to the more negative (0 volts). This may seem strange, as we know that current actually consists of electrons migrating from a negative source to a positive target. The reason for this inconsistency is that the existence of electricity was discovered long before it was fully understood. Electro-plated artifacts have been discovered in Egyptian pyramids, but it wasn't until the early part of the 20th century that George Thomson proved the existence of the electron at the University of Aberdeen, Scotland. The men who established the original electrical theories had to make decisions about things they didn't fully understand. The direction of current flow is one such example; for a variety of reasons it was originally believed that current flowed from positive to negative. As you may imagine, this inconsistency can, and does, cause endless problems.

Capacitance

Now imagine a full water tank A connected by a blocked pipe to an empty water tank B. Assume that the contents of tank A are magically maintained at the same level regardless of the amount of water that is removed. At some time T_0, the pipe is unblocked and tank B starts to fill. By the time T_{FULL}, tank B will have reached the same level as tank A (Figure 3.6).

The speed with which tank B fills depends on the rate with which water flows between the two tanks. The rate of water flow depends on the difference in pressure between the two ends of the pipe and any resistance to the flow caused by the pipe. When the water starts to flow between the tanks at time T_0, there is a large pressure differential between the end of the pipe in tank A and the end in tank B; however, as tank B fills, the pressure

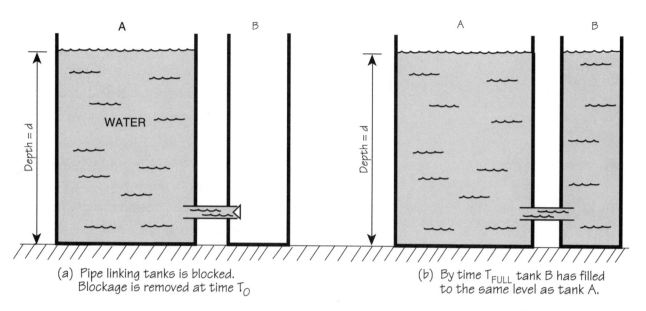

(a) Pipe linking tanks is blocked. Blockage is removed at time T_O

(b) By time T_{FULL} tank B has filled to the same level as tank A.

Figure 3.6: Water tank representation of capacitance

differential between the tanks is correspondingly reduced. This means that tank B fills faster at the beginning of the process than it does at the end. The rate at which tank B fills has an exponential characteristic best illustrated by a graph (Figure 3.7)

The electronic equivalent of tank B stores electrical charge. This ability to store charge is called *capacitance* measured in units of farads.

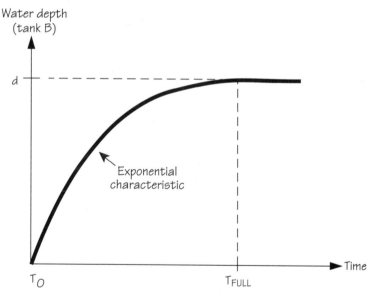

Figure 3.7: Rate at which water tank capacitor fills

Capacitances occur naturally in electronic circuits and engineers generally try to ensure that their values are as low as possible; however, in some cases designers may make use of special capacitive components called *capacitors*. One type of capacitor is formed from two metal plates separated by a layer of insulating material; the resulting capacitance (C) depends on the surface

Approximate actual size

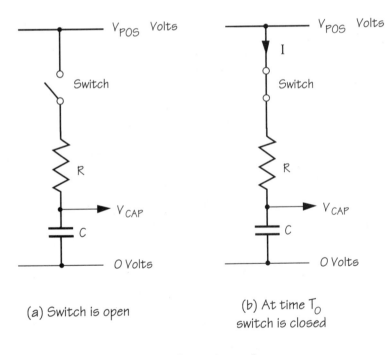

(a) Discrete component   (b) Symbol

**Figure 3.8: Capacitor:
component and symbol**

(a) Switch is open

(b) At time T_0
switch is closed

**Figure 3.9: Resistor-capacitor-
switch circuit**

area of the plates, the size of the gap between them, and the material used to fill the gap. Capacitors come in many shapes and sizes; an example could be as shown in Figure 3.8.

Now consider a simple circuit consisting of a resistor, a capacitor, and a switch. Initially the switch is in the open position, the capacitor voltage V_{CAP} is 0 volts, and no current is flowing (Figure 3.9).

When the switch is closed, any difference in potential between V_{POS} and V_{CAP} will cause current to flow through the resistor. As usual, the direction of current flow is illustrated in the classical rather than the actual sense. The current flowing through the resistor causes the capacitor to charge towards V_{POS}. But as the capacitor charges, the difference in voltage between V_{POS} and V_{CAP} decreases, and consequently so does the current (Figure 3.10).

The maximum current I_{MAX} occurs at time T_0 when there is the greatest difference between V_{POS} and V_{CAP}; from Ohm's law, $I_{MAX} = V_{POS}/R$. The capacitor is considered to be completely charged by time T_{FULL}, at which point the flow of current falls to zero.

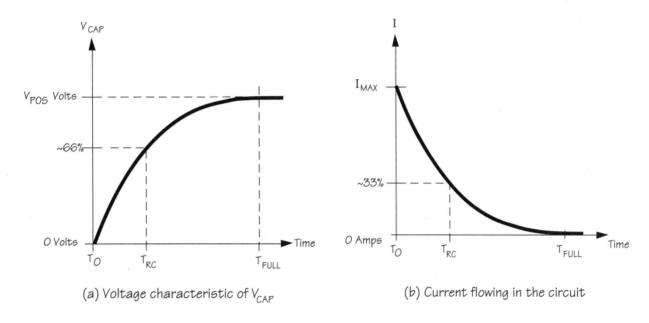

(a) Voltage characteristic of V_{CAP} (b) Current flowing in the circuit

Figure 3.10 Voltage and current characteristics of resistor-capacitor-switch circuit

The time T_{RC} equals $R \times C$ and is known as the *RC time constant*. With R in ohms and C in farads, the resulting T_{RC} is in units of seconds. The RC time constant is approximately equal to the time taken for V_{CAP} to achieve 66% of its final value and for the current to fall to 33% of its initial value.

Inductance

A difference in electrical potential between two ends of a conducting wire causes current to flow, and current flowing through a wire causes an electromagnetic field to be generated around the wire:

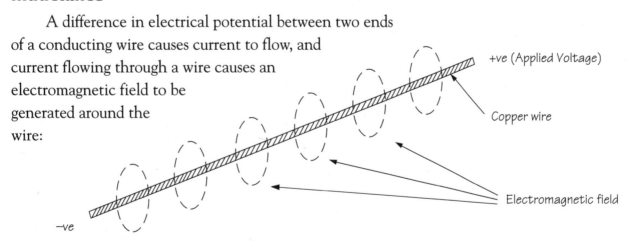

Figure 3.11: Current flow generates electromagnetic field

Correspondingly, if a piece of wire is moved through an externally generated electromagnetic field, it cuts the lines of electromagnetic flux, resulting in an electrical potential being generated between the two ends of the wire:

Figure 3.12: Conductor cutting electromagnetic field generates electrical potential

Engineers sometimes make use of components called *inductors*, which may be formed by winding a wire into a coil. When a current is passed through the coil, the result is an intense electromagnetic field:

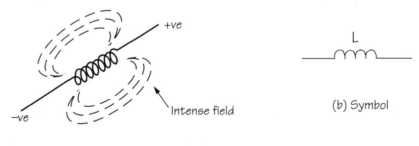

(a) Inductor Component

(b) Symbol

Figure 3.13: Inductor: component and symbol

Now consider a simple circuit consisting of a resistor, an inductor, and a switch. Initially the switch is in the open position, the inductor voltage V_{IND} is at V_{POS} volts, and no current is flowing (Figure 3.14).

As the inductor is formed from a piece of conducting wire, one might expect that closing the switch at time T_0 would immediately cause V_{IND} to drop to 0 volts; however, when the switch is closed and current begins to flow, the inductor's electromagnetic field starts to form. As the field grows in strength the lines of flux are pushed out from the center, and in the process they cut through the loops of wire forming the coil. This has the same effect as moving a conductor through an electromagnetic field and a voltage differential is created between the ends of the coil. This generated voltage is such that it attempts to oppose the changes causing it (Figure 3.15).

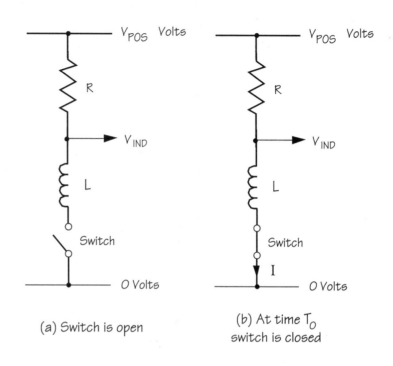

Figure 3.14: Resistor-inductor-switch circuit

This effect is called *inductance* and is measured in units of henries.[3] As time progresses, the coil's inductance is overcome and its electromagnetic field is fully established. Thus, by time T_{STABLE}, the inductor appears little different from any other piece of wire in the circuit (except for its concentrated electromagnetic field). This will remain the case until some new event occurs to disturb the circuit—for example, opening the switch again.

---

[3] Actually, the official unit of inductance is the henry, but it's difficult to pen a sentence without being obliged to use the plural (and good grief I tried). An inductor has an inductance of one henry when a potential of one volt is induced in it by a current changing at the rate of one amp per second.

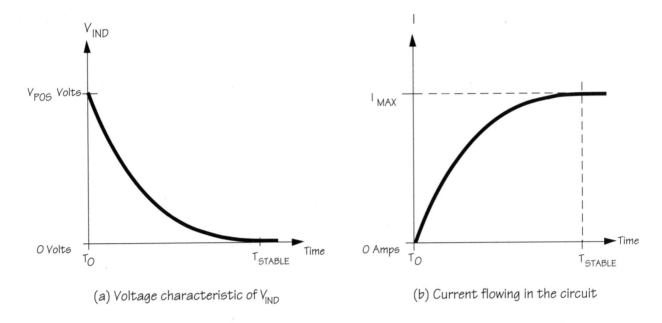

(a) Voltage characteristic of V_{IND}

(b) Current flowing in the circuit

**Figure 3.15: Voltage and current characteristics of
resistor-inductor-switch circuit**

Once a circuit has reached a steady-state condition where nothing is
changing, any capacitors act like open switches and any inductors act like
simple pieces of wire. The effects of these components only come into play
when signals are transitioning between different values.

Unit Qualifiers

Engineers often work with very large or very small values of voltage,
current, resistance, capacitance and inductance. As an alternative to writing
endless zeros, electrical quantities can be annotated with the qualifiers
given in Table 3.1. For example, 15 megaohms (15 MΩ)[4] means fifteen
million ohms, 4 milliamps (4 mA) means four thousandths of an amp, and
20 femtofarads (20 fF) means a very small capacitance indeed.

---

[4] The Greek letter omega, Ω, is often used to represent ohms.

Table 3.1

Name	Symbol	Value	Meaning
peta	P	10^{15}	One thousand million million
tera	T	10^{12}	One million million
giga	G	10^{9}	One thousand million
mega	M	10^{6}	One million
kilo	K	10^{3}	One thousand
milli	m	10^{-3}	One thousandth
micro	μ	10^{-6}	One millionth
nano	n	10^{-9}	One thousandth of one millionth
pico	p	10^{-12}	One millionth of one millionth
femto	f	10^{-15}	One thousandth of one millionth of one millionth
atlo	a	10^{-18}	One millionth of one millionth of one millionth

Semiconductors: Diodes and Transistors

Most materials are either conductors, insulators, or somewhere in between, but a special class of materials known as *semiconductors* can exhibit both conducting and insulating properties. The first semiconductor to undergo evaluation was the element germanium (chemical symbol Ge). However, for a variety of reasons, silicon (chemical symbol Si) replaced germanium as the semiconductor of choice. As silicon is the main consti-tuent of sand and one of the most common elements on earth (silicon accounts for approximately 28% of the earth's crust), we aren't in any danger of running out of it in the foreseeable future.

Pure crystalline silicon acts as an insulator; however, scientists at Bell Laboratories in the United States found that, by inserting certain impurities into the crystal lattice, they could make silicon act as a conductor. The process of inserting the impurities is known as *doping*, and the most com-monly used dopants are boron atoms with three electrons in their outermost electron shells and phosphorus atoms with five.

If a pure piece of silicon is surrounded by a gas containing boron or phosphorus and heated in a high temperature oven, the boron or phosphorus atoms will permeate the crystal lattice and displace some silicon atoms without disturbing other atoms in the vicinity. This process is known as *diffusion*. Boron-doped silicon is called *P-type* silicon and phosphorus-doped silicon is called *N-type* (Figure 4.1).

Figure 4.1: Creating P-type and N-type silicon

Because boron atoms only have three electrons in their outermost electron shells, they can only make bonds with three of the silicon atoms surrounding them. Thus, the site of a boron atom in the silicon crystal will accept a free electron with relative ease and is therefore known as an *acceptor*. Similarly, because phosphorous atoms have five electrons in their outermost electron shells, the site of a phosphorous atom in the silicon crystal will donate an electron with relative ease and is therefore known as a *donor*.

Diodes

Individually, both P-type and N-type silicon are reasonably good conductors; however, interesting things happen if a piece of silicon is doped such that part is P-type and part is N-type:

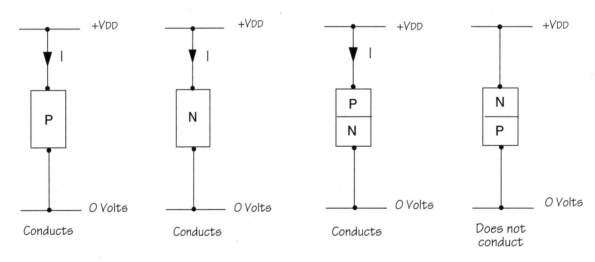

Figure 4.2: Comparison of pure and mixed P-type and N-type silicon

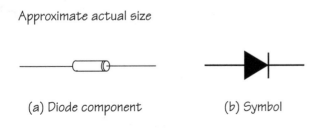

(a) Diode component   (b) Symbol

Figure 4.3: Diode: component and symbol

The silicon with both P-type and N-type material only conducts electricity in one direction; in the other direction it behaves like an open switch. These structures, known as *diodes*, come in many shapes and sizes; an example could be as shown in Figure 4.3.

If the triangular body of the symbol is pointing in the classical direction of current flow (more positive to more negative) the diode will conduct. An individually packaged diode consists of a piece of silicon, with connections to external leads, all encapsulated in a protective package (the silicon is typically smaller than a grain of sand). The package protects the silicon from moisture and other impurities and helps to conduct heat away from the silicon when the diode is operating.

Bipolar Junction Transistors

More complex components called *transistors* can be created by forming a sandwich out of three regions of doped silicon. One family of transistors are known as bipolar junction transistors (BJTs) of which there are two basic types called *NPN* and *PNP*; these names relate to the way in which the silicon is doped:

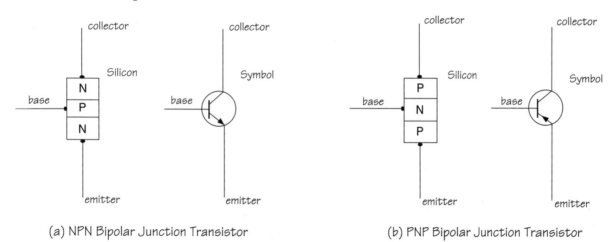

Figure 4.4: Bipolar junction transistors

In the analog world a transistor can be used as a voltage amplifier, a current amplifier, or a switch; in the digital world a transistor is primarily considered to be a switch. The structure of a transistor between the *collector* and *emitter* terminals is similar to two diodes connected back-to-back. Two diodes connected in this way would typically not conduct; however, by applying signals to the *base* terminal the transistor can be turned ON or OFF. If the transistor is turned ON, it acts like a closed switch and allows current to flow between the collector and the emitter; if the transistor is turned OFF, it acts like an open switch and no current flows. We may think of the collector and emitter as *data* terminals, and the base as the *control* terminal.

As for a diode, an individually packaged transistor consists of the silicon, with connections to external leads, all encapsulated in a protective package (the silicon is typically smaller than a grain of sand). The package protects the silicon from moisture and other impurities and helps to conduct heat away from the silicon when the transistor is operating.

Metal-Oxide Semiconductor Field-Effect Transistors

Another family of transistors are known as metal-oxide semiconductor field-effect transistors (MOS FETs) of which there are two basic types called *NMOS* and *PMOS*; once again these names relate to the way in which the silicon is doped (Figure 4.5).

In this case the *drain* and *source* form the *data* terminals and the *gate* acts as the *control* terminal. Unlike bipolar devices, the control terminal is connected to a conducting plate which is insulated from the silicon by a layer of nonconducting oxide. In the original devices the conducting plate was metal—hence the term *metal-oxide*. When a signal is applied to the gate terminal, the plate, insulated by the oxide, creates an electromagnetic field which turns the transistor ON or OFF—hence the term *field-effect*.

(a) NMOS Field-Effect Transistor   (b) PMOS Field-Effect Transistor

Figure 4.5: Metal-oxide semiconductor field-effect transistors

This book concentrates on metal-oxide semiconductor transistors, because their symbols, construction, and operation are easier to understand than those of bipolar junction transistors.

The Transistor as a Switch

To illustrate the application of a transistor as a switch, first consider a simple circuit comprising a resistor and a real switch:

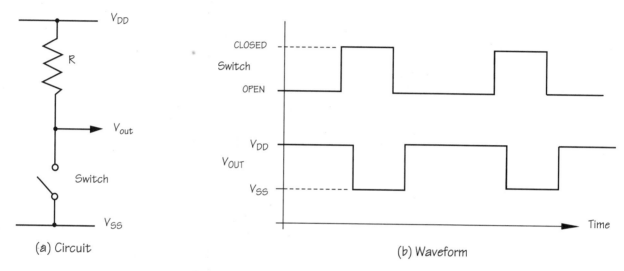

(a) Circuit   (b) Waveform

Figure 4.6: Resistor-switch circuit

The labels V_{DD} and V_{SS} are commonly used in circuits employing metal-oxide semiconductor transistors. At this point we have little interest in their actual values and, for the purpose of these examples, need only assume that V_{DD} is more positive than V_{SS}.

When the switch is OPEN, V_{OUT} is connected via the resistor to V_{DD}; when the switch is CLOSED, V_{OUT} is connected via the switch directly to V_{SS}. In this latter case, V_{OUT} takes the value V_{SS} because, like people, electricity takes the path of least resistance, and the resistance to V_{SS} through the closed switch is far less than the resistance to V_{DD} through the resistor. The waveforms in the illustration above show a delay between the switch operating and V_{OUT} responding. Although this delay is extremely small, it is important to note that there will always be some element of delay in any physical system.

Now consider the case where the switch is replaced with an NMOS transistor whose control input can be switched between V_{DD} and V_{SS}:

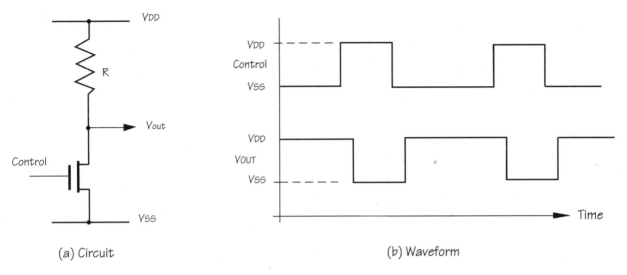

(a) Circuit   (b) Waveform

Figure 4.7: Resistor-NMOS transistor circuit

When the control input to an NMOS transistor is connected to V_{SS}, the transistor is turned OFF and acts like an open switch; when the control input is connected to V_{DD}, the transistor is turned ON and acts like a closed switch. Thus, the transistor functions in a similar manner to the switch. However, a switch is controlled by hand and can only be operated a few times a second, but a transistor's control input can be driven by other transistors allowing it to be operated millions of times a second.

Gallium Arsenide Semiconductors

Silicon is known as a 4-valence semiconductor because it has four electrons available to make bonds in its outermost electron shell. Although silicon is the most commonly used semiconductor, there is another that requires some mention. The element gallium (chemical symbol Ga) has three electrons available in its outermost shell and the element arsenic (chemical symbol As) has five. A crystalline structure of gallium arsenide (GaAs) is known as a 3:5-valence semiconductor and can be doped with impurities in a similar manner to silicon.

In a number of respects GaAs is preferable to silicon; GaAs transistors are approximately eight times faster and use one-tenth of the power of their silicon equivalents. However, pure GaAs is difficult to manufacture, hard to work with, and somewhat fragile.

Primitive Logic Functions

Consider an electrical circuit consisting of a power supply, a light, and two switches connected in *series* (one after the other); the switches are the inputs to the circuit and the light is the output. A *truth table* provides a convenient way to represent the operation of the circuit:

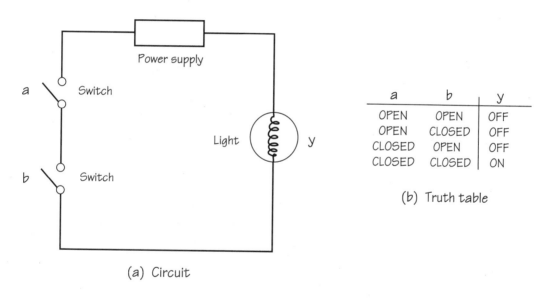

a	b	y
OPEN	OPEN	OFF
OPEN	CLOSED	OFF
CLOSED	OPEN	OFF
CLOSED	CLOSED	ON

(b) Truth table

(a) Circuit

Figure 5.1: Switch representation of a 2-input AND function

As the light is only ON when both the *a* <u>and</u> *b* switches are CLOSED, this circuit could be said to perform a 2-input AND function.[1] In fact, the results depend on the way in which the switches are connected; consider another circuit in which two switches are connected in *parallel* (side by side) (Figure 5.2).

---

[1] A 3-input version could be constructed by adding a third switch in series with the first two.

a	b	y
OPEN	OPEN	OFF
OPEN	CLOSED	ON
CLOSED	OPEN	ON
CLOSED	CLOSED	ON

(b) Truth table

(a) Circuit

Figure 5.2: Switch representation of a 2-input OR function

As the light is ON when either a <u>or</u> b are CLOSED, this circuit could be said to provide a 2-input OR function.[2] In a limited respect, we might consider that these circuits are making simple logical decisions; two switches offer four combinations of OPEN and CLOSED, but only certain combinations cause the light to be turned ON.

Logic functions such as AND and OR are generic concepts that can be implemented in a variety of ways, including switches as illustrated above transistors for use in computers, and even pneumatic devices for use in hostile environments such as steel works or nuclear reactors. Thus, instead of drawing circuits using light switches, it is preferable to make use of more abstract forms of representation. This permits designers to specify the function of systems with minimal consideration as to their final physical realization. To facilitate this, special symbols are employed to represent logic functions, and truth table assignments are specified using the abstract terms FALSE and TRUE. This is because assignments such as OPEN, CLOSED, ON, and OFF may imply a particular implementation.

---

[2] A 3-input version could be constructed by adding a third switch in parallel with the first two.

BUF and NOT Functions

The simplest of all the logic functions are known as BUF and NOT:

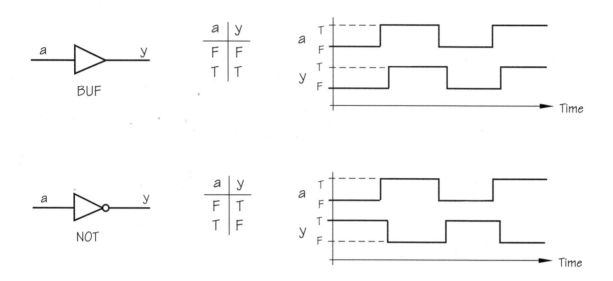

Figure 5.3: BUF and NOT functions

The F and T values in the truth tables are shorthand for FALSE and TRUE, respectively. The output of the BUF function has the same value as the input to the function; if the input is FALSE the output is FALSE, and if the input is TRUE the output is TRUE. By comparison, the small circle, or *bobble*,[3] on the output of the NOT symbol indicates an inverting function; if the input is FALSE the output is TRUE, and if the input is TRUE the output is FALSE.

As a reminder that these functions will eventually have physical realizations, the waveforms show delays between transitions on the inputs and corresponding responses at the outputs. The actual values of these delays depend on the technology used to implement the functions, but it is important to note that in any physical implementation there will always be *some* element of delay.

---

[3] Some engineers use the term *bobble*, others say *bubble*, and the rest try to avoid mentioning it at all.

Now consider the effect of connecting two NOT functions in series:

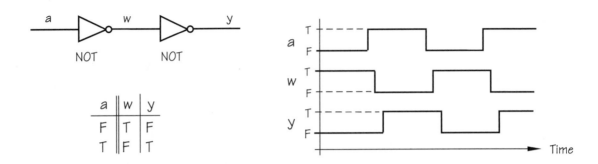

Figure 5.4: Two NOT functions connected in series

The first NOT gate inverts the value from the input, and the second NOT gate inverts it back again; thus, two negations cancel each other out. The end result is equivalent to that of a BUF function, except that each NOT function contributes an element of delay.

AND, OR, and XOR Functions

Three slightly more complex functions are known as AND, OR, and XOR (Figure 5.5). These AND and OR representations are the abstract equivalents of our original switch examples. In the case of the AND, the output is only TRUE if both a <u>and</u> b are TRUE; in the case of the OR, the output is TRUE if either a <u>or</u> b are TRUE. In fact, the OR should more properly be called an *inclusive-OR*, because the TRUE output cases include the case when both inputs are TRUE. Contrast this with the *exclusive-OR*, or XOR, where the TRUE output cases exclude the case when both inputs are TRUE.

NAND, NOR, and XNOR Functions

Now, consider the effect of a NOT function appended to the output of the AND function (Figure 5.6). This combination of functions occurs frequently in designs; similarly, the outputs of the OR and XOR functions are often inverted with NOT functions.

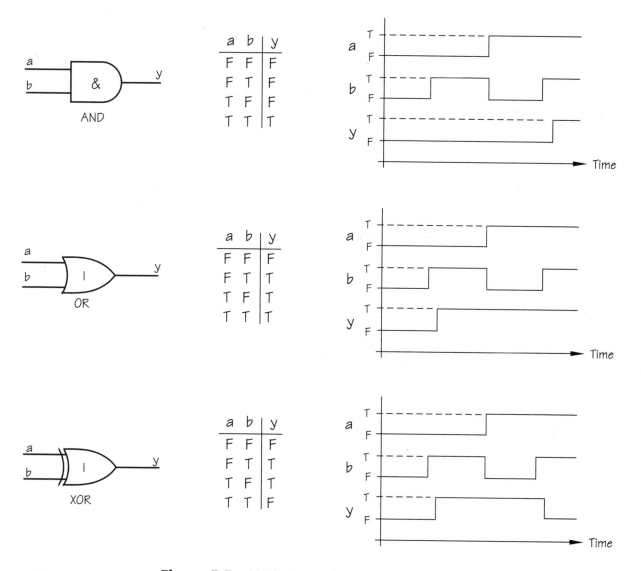

Figure 5.5: AND, OR and XOR functions

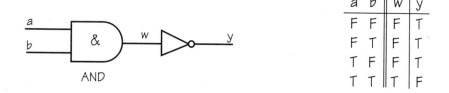

Figure 5.6: AND function followed by NOT function

This leads to three more primitive functions called NAND (NOT-AND), NOR (NOT-OR) and NXOR (NOT-XOR)[4] (Figure 5.7).

The bobbles on the outputs indicate inverting functions; one way to visualize this is that the symbol for the NOT function has been forced back into the preceding symbol until only the bobble remains visible.

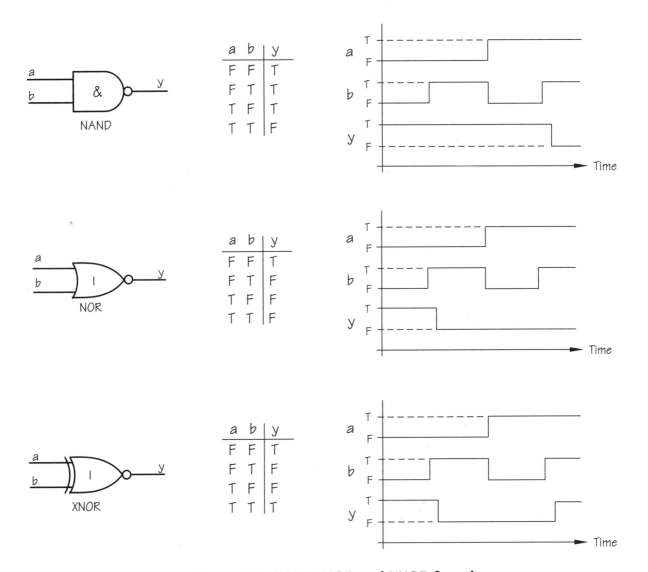

Figure 5.7: NAND, NOR and XNOR functions

---

[4] In practice, the NXOR is more commonly referred to as an XNOR (exclusive-NOR).

Functions versus Gates

Simple functions such as BUF, NOT, AND, NAND, OR, NOR, XOR, and XNOR are often known as *primitive gates*, *primitives*, or simply *gates*. Strictly speaking, the term *logic function* implies an abstract mathematical relationship, while *logic gate* implies an underlying physical implementation. In practice these terms are often used interchangeably.

More complex functions can be constructed by combining primitive gates in different combinations. A complete design usually employs a great many logic functions connected together to achieve the required result. When the time arrives to translate the abstract representation into a particular physical implementation, the logic symbols are converted into appropriate equivalents such as switches, transistors, or pneumatic valves. Similarly, the FALSE and TRUE logic values are mapped into appropriate equivalents such as switch positions, voltage levels, or air pressures. The majority of designs are translated into a single technology; however, one of the advantages of abstract representations is that they allow designers to implement different portions of a single design in dissimilar technologies with relative ease. Throughout the remainder of this book we will be concentrating on electronic implementations.

Finally, if some of the above seemed to be a little esoteric, consider two light switches controlling a light in your hallway at home—if both of the switches are UP or DOWN the light is ON; for any other combination the light is OFF. Constructing a truth table reveals a classic example of an XNOR function.

Using Transistors to Build Primitive Logic Functions

There are several different families of transistors available to designers and, although the actual implementations vary, each can be used to construct primitive logic gates. This book concentrates on the metal-oxide semiconductor transistors introduced in Chapter 4, because their symbols, construction, and operation are easier to understand than are bipolar junction transistors.

Logic gates can be created using only NMOS or only PMOS transistors; however, a popular implementation called *complementary metal-oxide semiconductor* (CMOS) makes use of both NMOS and PMOS transistors connected in a complementary manner.

CMOS gates operate from two voltage levels, which are usually given the labels V_{DD} and V_{SS}. To some extent the actual values of V_{DD} and V_{SS} are irrelevant as long as V_{DD} is sufficiently more positive than V_{SS}. There are also two conventions known as *positive logic and negative logic*.[1] Under the positive logic convention used throughout this book, the more positive V_{DD} is assigned the value of logic 1, and the more negative V_{SS} is assigned the value of logic 0. In Chapter 5 it was noted that truth table assignments can be specified using the abstract values FALSE and TRUE. However, for reasons that are more fully examined in Appendix B, electronic designers usually represent FALSE and TRUE as 0 and 1, respectively.

---

[1] The positive-logic and negative-logic conventions are discussed in detail in Appendix B.

NOT and BUF Gates

The simplest logic function to implement in CMOS is a NOT gate:

a	y
0	1
1	0

Figure 6.1: CMOS implementation of a NOT gate

The small circle, or bobble, on the control input of transistor Tr_1 indicates a PMOS transistor. The bobble is used to indicate that this transistor has an *active-low* control; a logic 0 applied to the control input turns the transistor ON and a logic 1 turns it OFF. The lack of a bobble on the control input of transistor Tr_2 indicates an NMOS transistor. The lack of a bobble says that this transistor has an *active-high* control; a logic 1 applied to the control input turns the transistor ON and a logic 0 turns it OFF.

Thus, when a logic 0 is applied to input *a*, transistor Tr_1 is turned ON, transistor Tr_2 is turned OFF, and output *y* is connected to logic 1 via Tr_1. Similarly, when a logic 1 is applied to input *a*, transistor Tr_1 is turned OFF, transistor Tr_2 is turned ON, and output *y* is connected to logic 0 via Tr_2.

Don't worry if all this seems a bit confusing at first. The main points to remember are that a logic 0 turns the PMOS transistor ON and the NMOS transistor OFF, while a logic 1 turns the PMOS transistor OFF and the NMOS transistor ON. It may help to visualize the NOT gate's operation in terms of switches rather than transistors:

a	y
0	1
1	0

Figure 6.2: NOT gate's operation represented in terms of switches

Surprisingly, a non-inverting BUF gate is more complex than an inverting NOT gate—a BUF gate requires four transistors because it is constructed from two NOT gates connected in series:

Figure 6.3: CMOS implementation of a BUF gate

The first NOT gate is formed from transistors Tr_1 and Tr_2, while the second is formed from transistors Tr_3 and Tr_4. A logic 0 applied to input a is inverted to a logic 1 on w, and then inverted back again to a logic 0 on output y. Similarly, a logic 1 on a is inverted to a logic 0 on w, and then inverted back again to a logic 1 on y.

Around this stage it is not unreasonable to question the need for BUF gates in the first place—after all, their logical function could be achieved using a simple piece of wire. In fact, BUF gates may be used for a number of reasons: to isolate signals, to provide increased drive capability, or to add an element of delay.

NAND and AND Gates

The implementations of the NOT and BUF gates shown above illustrate an important point—it is generally easier to implement an inverting function than its non-inverting equivalent. In the same way that a NOT is easier to implement than a BUF, a NAND is easier to implement than an AND, and a NOR is easier to implement than an OR. More significantly, inverting functions typically require fewer transistors and operate faster than their non-inverting counterparts. This can obviously be an important design consideration. Consider a 2-input NAND gate which requires four transistors[2] (Figure 6.4).

---

[2] A 3-input version could be constructed by adding an additional PMOS transistor in parallel with Tr_1 and Tr_2, and an additional NMOS transistor in series with Tr_3 and Tr_4.

Figure 6.4: CMOS implementation of a 2-input NAND gate

When both a and b are set to logic 1, transistors Tr_1 and Tr_2 are turned OFF, transistors Tr_3 and Tr_4 are turned ON, and output y is connected to logic 0 via Tr_3 and Tr_4. Any other combination of inputs results in one or both of Tr_3 and Tr_4 being turned OFF, one or both of Tr_1 and Tr_2 being turned ON, and output y being connected to logic 1 via Tr_1 and/or Tr_2. Once again, it may help to visualize the gate's operation in terms of switches rather than transistors:

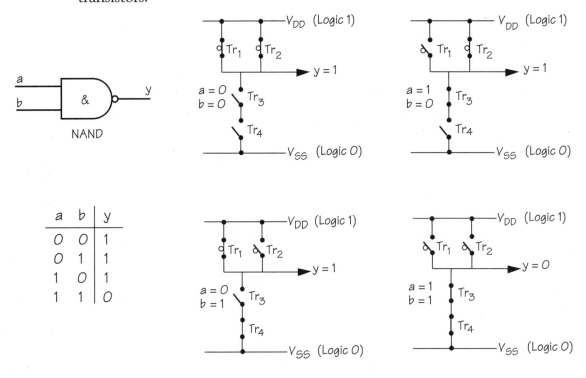

Figure 6.5: NAND gate's operation represented in terms of switches

Now consider an AND gate which is formed by inverting the output of a NAND with a NOT; thus, a 2-input AND requires six transistors:[3]

a	b	y
0	0	0
0	1	0
1	0	0
1	1	1

Figure 6.6: CMOS implementation of a 2-input AND gate

NOR and OR Gates

A similar story occurs in the case of NOR gates and OR gates. First, consider a 2-input NOR which requires four transistors:[4]

a	b	y
0	0	1
0	1	0
1	0	0
1	1	0

Figure 6.7: CMOS implementation of a 2-input NOR gate

When both a and b are set to logic 0, transistors Tr_3 and Tr_4 are turned OFF, transistors Tr_1 and Tr_2 are turned ON, and output y is connected to logic 1 via Tr_1 and Tr_2. Any other combination of inputs results in one or both of Tr_1 and Tr_2 being turned OFF, one or both of Tr_3 and Tr_4 being turned ON, and output y being connected to logic 0 via Tr_3 and/or Tr_4.

---

[3] Remember that electronic designers are cunning little devils with lots of tricks up their sleeves. In fact, it's possible to create an AND gate using only one transistor and a resistor (see the discussion on pass-transistor logic later in this chapter).

[4] A 3-input version could be constructed by adding an additional PMOS transistor in series with Tr_1 and Tr_2, and an additional NMOS transistor in parallel with Tr_3 and Tr_4.

Once again, an OR gate is formed by inverting the output of a NOR with a NOT; thus a 2-input OR requires six transistors:

a	b	y
0	0	0
0	1	1
1	0	1
1	1	1

Figure 6.8: CMOS implementation of a 2-input OR gate

XNOR and XOR Gates

The concepts of NAND, AND, NOR, and OR are relatively easy to understand because they map onto the way we think in everyday life. For example, a textual equivalent of a NOR could be: "If it's windy *or* if it's raining then I'm *not* going out."

By comparison, the concepts of XOR and XNOR can be a little harder to grasp because we don't usually consider things in these terms. A textual equivalent of an XOR could be: "If it *is* windy *and* it's *not* raining, *or* if it's *not* windy *and* it *is* raining, then I *will* go out." Although this does make sense, we don't often find ourselves making decisions in this way.

For this reason, it is natural to assume that XNOR and XOR gates would be a little more difficult to construct. However, these gates are full of surprises, both in the way that they work and the purposes for which they can be used. For example, a 2-input XNOR can be implemented using only four transistors:[5]

a	b	y
0	0	1
0	1	0
1	0	0
1	1	1

Figure 6.9: CMOS implementation of a 2-input XNOR gate

[5] Note that there are no such beasts as XNOR or XOR *primitives* with more than two inputs; however, equivalent functions with more than two inputs can be constructed by connecting a number of 2-input primitives together.

The NOT gate would be constructed in the standard way using two transistors as described above, but the XNOR differs from the previous gates in the way that transistors Tr_3 and Tr_4 are utilized. First, consider the case where input b is presented with a logic 0: transistor Tr_4 is turned OFF, transistor Tr_3 is turned ON, and output y is connected to the output of the NOT gate via Tr_3. Thus, when input b is logic 0, output y has the inverse of the value on input a. Now consider the case where input b is presented with a logic 1: transistor Tr_3 is turned OFF, transistor Tr_4 is turned ON, and output y is connected to input a via Tr_4. Thus, when input b is logic 1, output y has the same value as input a.

Unlike the other complementary gates, it is not necessary to invert the output of the XNOR to form an XOR. A little judicious rearranging of the components results in a 2-input XOR which also requires only four transistors:

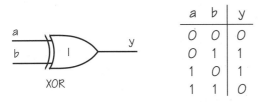

a	b	y
0	0	0
0	1	1
1	0	1
1	1	0

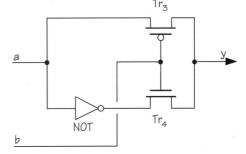

Figure 6.10: CMOS implementation of a 2-input XOR gate

First, consider the case where input b is presented with a logic 0: transistor Tr_4 is turned OFF, transistor Tr_3 is turned ON, and output y is connected to input a via Tr_3. Thus, when input b is logic 0, output y has the same value as input a. Now consider the case where input b is presented with a logic 1: transistor Tr_3 is turned OFF, transistor Tr_4 is turned ON, and output y is connected to the output of the NOT gate via Tr_4. Thus, when input b is logic 1, output y has the inverse of the value on input a.

Pass-Transistor Logic

In the BUF, NOT, AND, NAND, OR, and NOR gates shown above, the input signals and internal data signals are only used to drive control terminals on the transistors. By comparison, transistors Tr_3 and Tr_4 in the XOR and XNOR gates are connected so that input and internal data signals pass between their data terminals. This technique is known as *pass-transistor logic*. It can be attractive in that it minimizes the number of transistors required to implement a function, but it's not necessarily the best approach. Strange and unexpected effects can ensue if you're not careful and you don't know what you're doing.

An alternative solution for an XOR is to invert the output of the XNOR shown above with a NOT. Similarly, an XNOR can be constructed by inverting the output of the XOR shown above with a NOT. Although these new implementations now require six transistors rather than four, they are more robust because the NOT gates buffer the outputs and provide a higher drive capability. In many cases, XORs and XNORs are constructed from combinations of the other primitive gates. This increases the transistor count still further, but once again results in more robust solutions.

Having said all this, pass-transistor logic can be applicable in certain situations for designers who do know what they're doing. In the discussions above, it was noted that it is possible to create an AND using a single transistor and a resistor. Similarly, it's possible to create an OR using a single transistor and a resistor, and to create an XOR or an XNOR using only two transistors and a resistor. If you're feeling brave you can try to work out how to achieve these minimal implementations (solutions are given in Appendix G).

Chapter 7

Alternative Numbering Systems

The oldest known objects used to represent numbers are bones with notches discovered in western Europe. These bones date from the Aurignacian period 20,000 to 30,000 years ago and correspond to the first appearance of Cro-Magnon man. Of special interest, a wolf bone more than 20,000 years old with 55 notches in groups of five was discovered at Vestonice, Czechoslovakia in 1937. This is the first evidence of the tally system, which is still used occasionally to the present day and could therefore qualify as one of the most enduring human inventions.

Decimal (Base-10)

The commonly used decimal numbering system is based on ten digits: 0, 1, 2, 3, 4, 5, 6, 7, 8 and 9. The name decimal comes from the Latin *decem*, meaning ten. The symbols used to represent these digits arrived in Europe around the thirteenth century from the Arabs, who in turn borrowed them from the Hindus (and never gave them back). As the decimal system is based on ten digits, it is said to be *base-10* or *radix-10*.

With the exception of specialist requirements such as computing, base-10 numbering systems have been adopted almost universally—this is almost certainly due to the fact that humans happen to have ten fingers.[1] If mother nature had dictated six fingers on each hand, the outcome would most probably have been the common usage of base-12 numbering systems.

The decimal system is a *place-value* system, which means that the value of a particular digit depends both on the digit itself and its position within the number (Figure 7.1).

---

[1] Including thumbs.

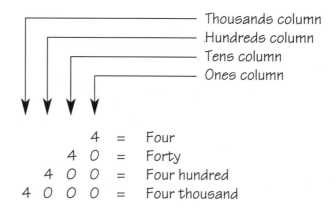

			4	=	Four
		4	0	=	Forty
	4	0	0	=	Four hundred
4	0	0	0	=	Four thousand

Figure 7.1: Place-value number systems

Every column in a place-value number has a *weight* associated with it, and each digit is combined with its column's weight to determine the final value of the number:

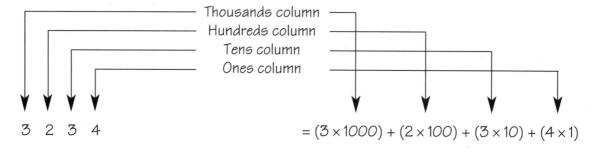

Figure 7.2: Combining digits with column weights in decimal

Counting in decimal commences at 0 and progresses up to 9, at which point all the available digits have been used. Thus, the next count causes the first column to be reset to 0 and the second column to be incremented resulting in 10. Similarly, when the count reaches 99, the next count causes the first column to be reset to zero and the second column to be incremented. But, as the second column already contains a 9, this causes *it* to be reset to 0 and the third column to be incremented resulting in 100 (Figure 7.3).

Although base-10 systems are anatomically convenient, they have few other advantages to recommend them. In fact, depending on your point of view, almost any other base (with the possible exception of nine) would be as good as, or better than, base-10, which is only wholly divisible by 2 and 5. For many arithmetic operations, the use of a base that is wholly divisible by many numbers, especially the smaller values, conveys certain advantages. An educated layman may well prefer a base-12 system on the basis that 12 is wholly divisible by 2, 3, 4 and 6. For their own esoteric purposes, some mathematicians would ideally prefer a system with a prime number as a base; for example, seven or eleven.

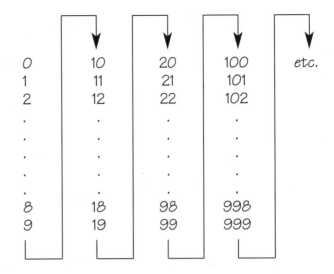

Figure 7.3: Counting in decimal

Duo-Decimal (Base-12)

Number systems with bases other than ten have sprouted up like weeds throughout history. Some cultures made use of duo-decimal (base-12) systems; instead of counting fingers they counted finger-joints. The thumb is used to point to the relevant joint and, as each finger has three joints, the remaining fingers can be used to represent values from 1 through 12 (Figure 7.4).

This system is particularly useful if one wishes to count up to twelve while still maintaining a free hand to throw a spear at someone whom, we may assume, is not a close friend.

This form of counting may explain why the ancient Sumerians, Assyrians, and Babylonians divided their days into twelve periods, six for day and six for night. The lengths of the periods were adjusted to the seasons (since the length of daylight compared to night time varies throughout the year), but were approximately equal to two of our hours. In fact, the Chinese use a form of this system to the present day (Figure 7.5).

Thumb is used to point to relevant joint

Figure 7.4: Using finger joints to count in duo-decimal

11:00pm ⇨ 1:00am	≡	Hour of the Rat
1:00am ⇨ 3:00am	≡	Hour of the Ox
3:00am ⇨ 5:00am	≡	Hour of the Tiger
5:00am ⇨ 7:00am	≡	Hour of the Hare
7:00am ⇨ 9:00am	≡	Hour of the Dragon
9:00am ⇨ 11:00am	≡	Hour of the Snake
11:00am ⇨ 1:00pm	≡	Hour of the Horse
1:00pm ⇨ 3:00pm	≡	Hour of the Ram
3:00pm ⇨ 5:00pm	≡	Hour of the Monkey
5:00pm ⇨ 7:00pm	≡	Hour of the Cock
7:00pm ⇨ 9:00pm	≡	Hour of the Dog
9:00pm ⇨ 11:00pm	≡	Hour of the Boar

Figure 7.5: The Chinese twelve-hour day

If a similar finger-joint counting strategy is applied to both hands, the counter can represent values from 1 through 24 (Figure 7.6).

This may explain why the ancient Egyptians divided their days into twenty-four periods, twelve for day and twelve for night. Once again, the lengths of the periods were adjusted to the seasons, but were approximately equal to one of our hours.

Figure 7.6: Using finger joints on both hands to count to twenty-four

Sexagesimal (Base-60)

The Sumerians and the Babylonians after them used sexagesimal (base-60) numbering systems. The Babylonian sexagesimal system, which appeared between 1900 BC to 1800 BC, is also credited as being the first known place-value number system. While there is no definite proof as to the origins of the sexagesimal base, it's possible that this was an extension of the finger-joint counting schemes discussed above (Figure 7.7).

The finger joints of the left hand are still used to represent the values one through twelve; however, instead of continuing directly with the finger joints of the right hand, the thumb and fingers on the right hand are used to keep track of each count of twelve. When all of the right hand digits are extended the count is sixty (5 × 12 = 60).

Although sixty may appear to be a large value to have as a base, it does convey certain advantages. Sixty is the smallest number that can be wholly divided by 2 through 6; in fact, sixty can be wholly divided by 2, 3, 4, 5, 6, 10, 12, 15, 20 and 30. In addition to using base sixty, the Babylonians also made use of six and ten as sub-bases.

The water clock, or Clepsydra, is an ancient instrument for measuring time which is believed to have originated in Egypt. These clocks consisted of a container with a small hole through which the water escaped. Units of time were marked on the side of the container or on the side of a bowl used to collect the water. The length of the units corresponding to day and night could be adjusted by varying the shape of the container; for example, by having the top wider than the bottom.

Figure 7.7: Using fingers to count to sixty

The Concepts of Zero and Negative Numbers

The concept of numbers like 1, 2, and 3 developed a long time before the concept of zero. In the original Babylonian system a zero was simply represented by a space; the decimal equivalent would be as follows:

"126904"	would be written as	"1269 4"
"102056"	would be written as	"1 2 56"
"160014"	would be written as	"16 14"

Figure 7.8: Representing zeros using spaces

It is easy to see how this can lead to a certain amount of confusion, especially when attempting to portray multiple zeros next to each other. The problems can only be exacerbated if one is using a sexagesimal system and writing on tablets of damp clay in a thunderstorm.

After more than 1500 years of potentially inaccurate calculations, the Babylonians finally began to use a special sign for zero, which first appeared around 300 BC. Some say that this was one of the most significant inventions in the history of mathematics. However, the Babylonians only used the zero symbol as a place holder to separate digits—they didn't have the concept of zero as an actual value. If we were to use the 'Δ' character to represent the Babylonian place holder, the decimal equivalent to the clay tablet accounting records of the time would read something like:

Original Amount	−	*Fish Distributed*	=	*Fish Remaining*
1Δ52 fish	−	45 fish to Steph	=	1ΔΔ7 fish
	−	2Δ fish to Max	=	987 fish
	−	4Δ7 fish to Lucie	=	58Δ fish
	−	176 fish to Drew	=	4Δ4 fish
	−	4Δ4 fish to Abby	=	"No fish left"

Figure 7.9: Representing zeros using a place-holder

The last entry in the *Fish Remaining* column is particularly revealing. Due to the fact that the Babylonian zero was only a place holder and not a value, the accounting records had to say "No fish left" rather than "Δ fish".

The use of zero as an actual value, along with the concept of negative numbers, first appeared in India around 600 AD. Although negative numbers may appear to be reasonably obvious to us, they were not well understood until modern times. As recently as the eighteenth century, the great mathematician Leonhard Euler (1707–1783) believed that negative numbers were greater than infinity, and it was common practice to ignore any negative results returned from equations on the basis that they were meaningless.

Vigesimal (Base-20)

The Mayans, Aztecs, and Celts developed vigesimal (base-20) systems by counting using both fingers and toes. The Eskimos of Greenland, the Tamanas of Venezuela, and the Ainu of northern Japan are three of the many other groups of people who also make use of vigesimal systems. For example, to say fifty-three, the Greenland Eskimos would use the expression *"Inup pinga-jugsane arkanek-pingasut,"* which translates as *"Of the third man, three on the first foot."*[2] This means that the first two men contribute twenty each (ten fingers and ten toes), and the third man contributes thirteen (ten fingers and three toes).

To this day we bear the legacies of almost every number system our ancestors experimented with. From the duo-decimal systems we have twenty-four hours in a day (2×12), twelve inches in a foot, and special words such as dozen (12) and gross (144). Similarly, the Chinese have twelve hours in a day and twenty-four seasons in a year. From the sexagesimal systems we have sixty seconds in a minute, sixty minutes in an hour, and 360 degrees in a circle.[3] From the base twenty systems we have special words like score (20), as in *"Four score and seven years ago, our fathers brought forth upon this continent a new nation . . ."* This all serves to illustrate that number systems with bases other than ten are not only possible, but positively abound throughout history.

---

[2] George Ifrah, From One to Zero (A Universal History of Numbers).

[3] The 360 degrees is derived from the product of the Babylonian's main base (sixty) and their sub-base (six);
$6 \times 60 = 360$.

Quinary (Base Five)

One system that is relatively easy to understand is quinary (base-5) which uses the digits 0, 1, 2, 3 and 4. This system is particularly interesting in that a quinary finger-counting scheme is still in use today by merchants in the Indian state of Maharashtra near Bombay.

As with any place-value system, each column in a quinary number has a weight associated with it, where the weights are derived from the base. Each digit is combined with its column's weight to determine the final value of the number:

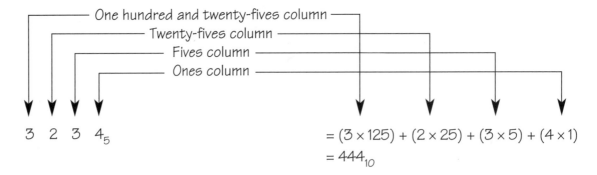

Figure 7.10: Combining digits with column weights in quinary

When using systems with bases other than ten, subscripts are used to indicate the relevant base; for example, $3234_5 = 444_{10}$ ($3234_{QUINARY} = 444_{DECIMAL}$). By convention, any value without a subscript is assumed to be in decimal.

Counting in quinary commences at 0 and progresses up to 4_5, at which point all the available digits have been used. Thus, the next count causes the first column to be reset to 0 and the second column to be incremented resulting in 10_5. Similarly, when the count reaches 44_5, the next count causes the first column to be reset to zero and the second column to be incremented. But, as the second column already contains a 4, this causes *it* to be reset to 0 and the third column to be incremented resulting in 100_5 (see Figure 7.11).

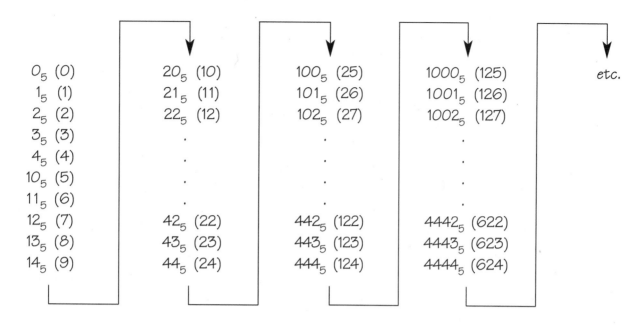

Figure 7.11: Counting in quinary

Binary (Base-2)

Digital systems are constructed out of logic gates which can only represent two states; thus, computers are obliged to make use of a number system comprising only two digits. Base-2 number systems are called binary and use the digits 0 and 1. As usual, each column in a binary number has a weight derived from the base, and each digit is combined with its column's weight to determine the final value of the number:

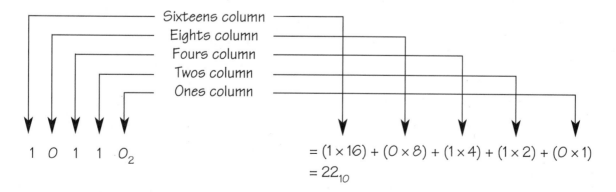

Figure 7.12: Combining digits with column weights in binary

Once again, subscripts are used to indicate the relevant base; for example, $10110_2 = 22_{10}$ ($10110_{BINARY} = 22_{DECIMAL}$). The term *binary digit* is usually abbreviated to *bit*; thus, the binary value 10110_2 would be said to be 5 bits wide. Additionally, a group of 4 bits is known as a *nybble* (sometimes called a *nibble*), and a group of 8 bits is known as a *byte*.[4]

Counting in binary commences at 0 and rather quickly progresses up to 1_2, at which point all the available digits have been used. Thus, the next count causes the first column to be reset to 0 and the second column to be incremented resulting in 10_2. Similarly, when the count reaches 11_2, the next count causes the first column to be reset to zero and the second column to be incremented. But, as the second column already contains a 1, this causes *it* to be reset to 0 and the third column to be incremented resulting in 100_2:

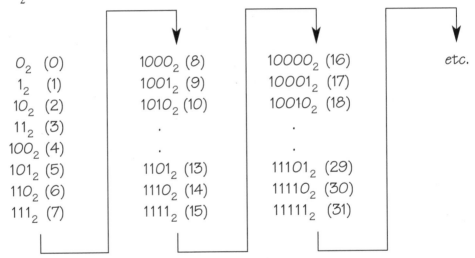

Figure 7.13: Counting in binary

Although binary mathematics is fairly simple, humans tend to find it difficult because the numbers are inclined to be long and laborious to manipulate. For example, the binary value 11010011_2 is relatively difficult to conceptualize, while its decimal equivalent 211_{10} is comparatively easy.

---

[4] So two nybbles make a byte! Engineers do have a sense of humor—it's just not very well developed.

Octal (Base-8) and Hexadecimal (Base-16)

Any number system having a base that is a power of two (2, 4, 8, 16, 32, etc.) can be easily mapped into its binary equivalent and vice versa. For this reason, electronics engineers typically make use of either the octal (base-8) or hexadecimal (base-16) systems.

As a base-16 system, hexadecimal requires 16 individual symbols to represent all of its digits. This poses something of a problem because there are only 10 Hindu-Arabic symbols available (0 through 9). One solution would be to create some new symbols, but some doubting Thomases regard this as less than optimal because it would necessitate the modification of existing typewriters and computer keyboards. As an alternative, the first six letters of the alphabet are brought into play:

Hexadecimal	0	1	2	3	4	5	6	7	8	9	A	B	C	D	E	F
Decimal	0	1	2	3	4	5	6	7	8	9	10	11	12	13	14	15

Figure 7.14: The 16 hexadecimal digits

The rules for counting in octal and hexadecimal are the same as for any other place-value system—when all the digits in a column are exhausted, the next count sets that column to zero and increments the column to the left (Figure 7.15).

Although not strictly necessary, binary, octal and hexadecimal numbers are often prefixed by leading zeros to pad them to whatever width is required. This padding is usually performed to give some indication as to the physical number of bits used to represent that value within a computer.

Each octal digit can be directly mapped onto three binary digits, and each hexadecimal digit can be directly mapped onto four binary digits (Figure 7.16).

In the original digital computers, data paths were often 9 bits, 12 bits, 18 bits, or 24 bits wide, which provides one reason for the original popularity of the octal system. Because each octal digit maps directly to three binary bits, these data-path values were easily represented in octal. More recently, digital computers have standardized on data-path widths that are integer

Decimal	Binary	Octal	Hexadecimal
0	00000000	000	00
1	00000001	001	01
2	00000010	002	02
3	00000011	003	03
4	00000100	004	04
5	00000101	005	05
6	00000110	006	06
7	00000111	007	07
8	00001000	010	08
9	00001001	011	09
10	00001010	012	0A
11	00001011	013	0B
12	00001100	014	0C
13	00001101	015	0D
14	00001110	016	0E
15	00001111	017	0F
16	00010000	020	10
17	00010001	021	11
18	00010010	022	12
:	:	:	:
etc.	etc.	etc.	etc.

Figure 7.15: Counting in octal and hexadecimal

Figure 7.16: Mapping octal and hexadecimal to binary

multiples of 8 bits—for example, 8 bits, 16 bits, 32 bits, or 64 bits. Because each hexadecimal digit maps directly to four binary bits, these data-path values are easily represented in hexadecimal. This may explain the decline in popularity of the octal system and the corresponding rise in popularity of the hexadecimal system.

Representing Numbers Using Powers

An alternative way of representing numbers is by means of *powers*—for example, 10^3, where 10 is the base value and the superscripted 3 is known as the power or exponent. We read 10^3 as *"Ten to the power of three."* The power specifies how many times the base value must be multiplied by itself; thus, 10^3 represents $10 \times 10 \times 10$. Any value can be used as a base:

Decimal (Base-10)			
10^0 = 1	=		1_{10}
10^1 = 10	=		10_{10}
10^2 = 10×10	=		100_{10}
10^3 = $10 \times 10 \times 10$	=		1000_{10}
Binary (Base-2)			
2^0 = 1	=	1_2	= 1_{10}
2^1 = 2	=	10_2	= 2_{10}
2^2 = 2×2	=	100_2	= 4_{10}
2^3 = $2 \times 2 \times 2$	=	1000_2	= 8_{10}
Octal (Base-8)			
8^0 = 1	=	1_8	= 1_{10}
8^1 = 8	=	10_8	= 8_{10}
8^2 = 8×8	=	100_8	= 64_{10}
8^3 = $8 \times 8 \times 8$	=	1000_8	= 512_{10}
Hexadecimal (Base-16)			
16^0 = 1	=	1_{16}	= 1_{10}
16^1 = 16	=	10_{16}	= 16_{10}
16^2 = 16×16	=	100_{16}	= 256_{10}
16^3 = $16 \times 16 \times 16$	=	1000_{16}	= 4096_{10}

Figure 7.17: Representing numbers using powers

Any base to the power of one is equal to itself; for example, $8^1 = 8$. Strictly speaking, a power of zero is not part of the series but, by convention, any base to the power of zero equals one; for example, $8^0 = 1$. Powers provide a convenient way to represent column-weights in place-value systems:

Decimal (Base-10)

Figure 7.18a: Using powers to represent column weights in decimal

Binary (Base-2)

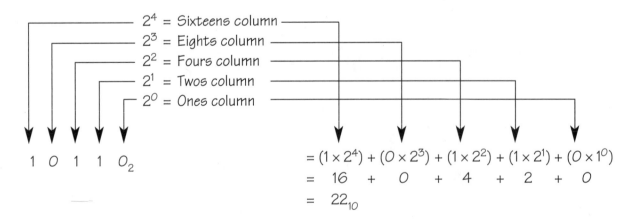

Figure 7.18b: Using powers to represent column weights in binary

Tertiary Logic

And finally, for reasons that become a little involved, communications theory tells us that optimal data transfer rates can be achieved if each data element represents three states. Now engineers wouldn't be the guys and gals they are if they weren't prepared to accept a challenge, and some experiments were performed with *tertiary logic*—logic gates based on three distinct voltage levels. The three voltages were used to represent the tertiary (base-3) values of 0, 1, and 2, and their logical equivalents FALSE, TRUE, and MAYBE.[5]

Quite apart from anything else, man has been interested in the properties of odd numbers since antiquity, often ascribing mystical and magical properties to them— for example, "lucky seven" and "unlucky thirteen." As Pliny the Elder (AD 23–79) said: *"Why is that we entertain the belief that for every purpose odd numbers are the most effectual?"*

However, while it's relatively easy to use transistors to generate two distinct voltage levels, it proved harder to generate a consistent intermediate voltage to represent a third value. Similarly, while it's relatively easy to use transistors to interpret (detect) two voltage levels, it's harder to interpret a third, intermediate value.[6] Additionally, creating and using an equivalent to Boolean algebra[7] that works with the three logic states FALSE, TRUE, and MAYBE is enough to make even the strongest among us quail. Thankfully, tertiary logic was abandoned; otherwise this book might have been substantially longer. Still, there's an old saying *"What goes around comes around,"* and it's not beyond the bounds of possibility that tertiary logic or an even bigger relative will rear its ugly head in the future.

---

[5] A tertiary digit is known as a "trit."

[6] Actually, it's not too difficult to generate and detect three voltage levels; the problem is to do it efficiently without using lots of transistors.

[7] Boolean algebra is introduced in Chapter 9.

Binary Arithmetic

Due to the fact that digital computers are constructed from logic gates which can represent only two states, they are obliged to make use of the binary number system which provides only two digits, 0 and 1. Unlike calculations on paper where both decimal and binary numbers can be of any size—limited only by the size of your paper, the endurance of your pencil, and your stamina—the numbers manipulated within a computer have to be mapped onto a physical system of logic gates and wires. Thus, the maximum value of a number inside a computer is dictated by the width of its data path—that is, the number of bits used to represent that number.

Unsigned Binary Numbers

Unsigned binary numbers can only be used to represent positive values. Consider the range of numbers that can be represented using 8 bits (Figure 8.1). Each 'x' character represents a single bit; the right-hand bit is known as the *least-significant bit (LSB)* because it represents the

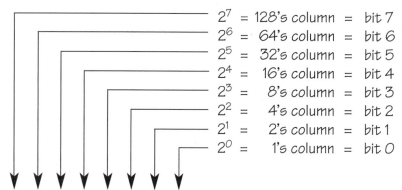

2^7	=	128's column	=	bit 7
2^6	=	64's column	=	bit 6
2^5	=	32's column	=	bit 5
2^4	=	16's column	=	bit 4
2^3	=	8's column	=	bit 3
2^2	=	4's column	=	bit 2
2^1	=	2's column	=	bit 1
2^0	=	1's column	=	bit 0

MSB → x x x x x x x x ←—LSB

$$0\ 0\ 0\ 0\ 0\ 0\ 0\ 0\ =\ 0_{10}$$
$$0\ 0\ 0\ 0\ 0\ 0\ 0\ 1\ =\ 1_{10}$$
$$0\ 0\ 0\ 0\ 0\ 0\ 1\ 0\ =\ 2_{10}$$
$$0\ 0\ 0\ 0\ 0\ 0\ 1\ 1\ =\ 3_{10}$$
$$0\ 0\ 0\ 0\ 0\ 1\ 0\ 0\ =\ 4_{10}$$

$$1\ 1\ 1\ 1\ 1\ 1\ 0\ 1\ =253_{10}$$
$$1\ 1\ 1\ 1\ 1\ 1\ 1\ 0\ =254_{10}$$
$$1\ 1\ 1\ 1\ 1\ 1\ 1\ 1\ =255_{10}$$

Figure 8.1: Unsigned binary numbers

smallest value, and the left hand bit is known as the *most-significant bit (MSB)* because it represents the largest value.

In computing it is usual to commence indexing from zero, so the least-significant bit is referenced as bit 0, and the most-significant bit (of an 8-bit value) is referenced as bit 7. Every bit can be individually assigned a value of 0 or 1, so a group of 8 bits can be assigned 2^8 (= 256) unique combinations of 0s and 1s. This means that an 8-bit unsigned binary number can be used to represent values in the range 0_{10} through $+255_{10}$.

Binary Addition

Two binary numbers may be added together using an identical process to that used for decimal addition. First, the two least-significant bits are added together to give a sum and, possibly, a carry-out to the next stage. This process is repeated for the remaining bits progressing towards the most-significant. For each of the remaining bits, there may be a carry-in from the previous stage and a carry-out to the next stage. To fully illustrate this process, consider the step-by-step addition of two 8-bit binary numbers (Figure 8.2).

The first carry-out occurs in case (*d*), where 1 + 1 = 2 or, in binary, $1_2 + 1_2 = 10_2$; thus, in this instance, the sum is 0 and there is a carry-out of 1 to the next stage. The second carry-out occurs in case (*e*), where the carry-in from the previous stage results in 1 + 1 + 1 = 3 or, in binary, $1_2 + 1_2 + 1_2 = 11_2$; thus, in this instance, the sum is 1 and there is a carry-out of 1 to the next stage. The third and final carry-out occurs in case (*f*).

Binary Subtraction

Unsigned binary numbers may be subtracted from each other using an identical process to that used for decimal subtraction. However, for reasons of efficiency, computers rarely perform subtractions in this manner; instead, these operations are typically performed by means of complement techniques.

There are two forms of complement associated with every number system, the *radix complement* and the *diminished radix complement,* where

```
    0  0  1  1  1  0  0 │1│           0  0  1  1  1  0 │0│ 1
 +  0  0  0  1  1  0  1 │0│        +  0  0  0  1  1  0 │1│ 0
 = ─────────────────────            = ────────────────────
                      1                               1  1
```

(a) Bit 0, $1 + 0 = 1_2$ (b) Bit 1, $0 + 1 = 1_2$

```
    0  0  1  1  1 │0│ 0  1           0  0  1  1 │1│ 0  0  1
 +  0  0  0  1  1 │0│ 1  0        +  0  0  0 ₁1 │1│ 0  1  0
 = ────────────────────            = ────────────────────
                0  1  1                            0  0  1  1
```

(c) Bit 2, $0 + 0 = 0_2$ (d) Bit 3, $1 + 1 = 10_2$

```
    0  0  1 │1│ 1  0  0  1           0  0 │1│ 1  1  0  0  1
 +  0  0 ₁0 │₁1│ 1  0  1  0       +  0 ₁0 │₁0│ ₁1 1  0  1  0
 = ────────────────────            = ────────────────────
             1  0  0  1  1                0  1  0  0  1  1
```

(e) Bit 4, $1 + 1 + carry\_in = 11_2$ (f) Bit 5, $1 + 0 + carry\_in = 10_2$

```
    0 │0│ 1  1  1  0  0  1        │0│ 0  1  1  1  0  0  1           57₁₀
 +  0 │₁0│ ₁0 ₁1 1  0  1  0     + │0│ ₁0 ₁0 ₁1 1  0  1  0        +  26₁₀
 = ────────────────────          = ─────────────────────          = ────
          1  0  1  0  0  1  1     = 0  1  0  1  0  0  1  1            83₁₀
```

(g) Bit 6, $0 + 0 + carry\_in = 1_2$ (h) Bit 7, $0 + 0 = 0_2$

Figure 8.2: Binary addition

radix refers to the base of the number system. Under the decimal (base-10) system, the radix complement is also known as the *tens complement* and the diminished radix complement is known as the *nines complement*. First, consider a decimal subtraction performed using the nines complement technique—a process known in ancient times as "*Casting out the nines*" (Figure 8.3).

Figure 8.3: Nines complement decimal subtraction

The standard way of performing the operation would be to subtract the subtrahend (283) from the minuend (647) which, as in this example, may require the use of one or more borrow operations. To perform the equivalent operation in nines complement, each of the digits of the subtrahend is first subtracted from a 9. The resulting nines complement value is added to the minuend, and then an end-around-carry operation is performed. The advantage of the nines complement technique is that it is never necessary to perform a borrow operation.

Now consider the same subtraction performed using the tens complement technique (Figure 8.4). The advantage of the tens complement is that it is not necessary to perform an end-around-carry; any carry-out resulting from the addition of the most-significant digits is simply dropped from the

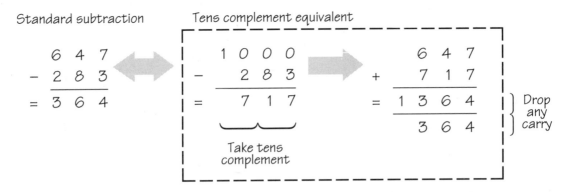

Figure 8.4: Tens complement decimal subtraction

final result. The disadvantage is that, during the process of creating the tens complement, it is necessary to perform a borrow operation for every digit in the subtrahend. This problem could be solved by first taking the nines complement of the subtrahend, adding one to the result, and then performing the remaining operations as for the tens complement.

Similar techniques may be employed with binary (base-2) numbers, where the radix complement is known as the *twos complement* and the diminished radix complement is known as the *ones complement*. First, consider a binary subtraction performed using the ones complement technique:

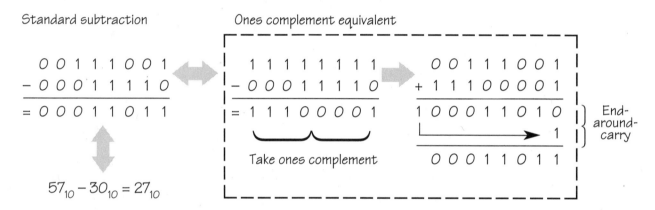

Figure 8.5: Ones complement binary subtraction

Once again, the standard way of performing the operation would be to subtract the subtrahend (00011110_2) from the minuend (00111001_2), which may require the use of one or more borrow operations. To perform the equivalent operation in ones complement, each of the digits of the subtrahend is first subtracted from a 1. The resulting ones complement value is added to the minuend, then an end-around-carry operation is performed. The advantage of the ones complement technique is that it is never necessary to perform a borrow operation. In fact, it isn't even necessary to perform a subtraction operation, because the ones complement of a binary number can be simply generated by inverting all of its bits; that is, by exchanging all the 0s with 1s and vice versa.

Now consider the same binary subtraction performed using the twos complement technique:

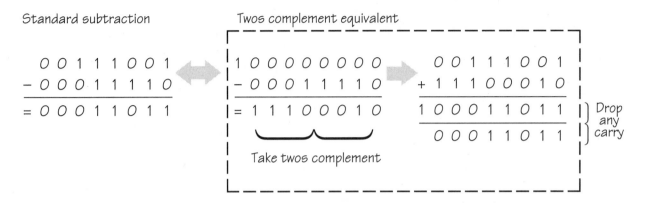

Figure 8.6: Twos complement binary subtraction

The advantage of the twos complement is that it is not necessary to perform an end-around-carry; any carry-out resulting from the addition of the two most-significant bits is simply dropped from the final result. The disadvantage is that, during the process of creating the twos complement, it is necessary to perform a borrow operation for almost every digit in the subtrahend. This problem can be solved by first taking the ones complement of the subtrahend, adding one to the result, and then performing the remaining operations as for the twos complement.

As fate would have it, there is a short-cut approach available to generate the twos complement of a binary number. Commencing with the least significant bit of the value to be complemented, each bit up to and including the first 1 is copied directly, and the remaining bits are inverted (Figure 8.7).

Unfortunately, all of the previous examples will return incorrect results if a larger value is subtracted from a smaller value; that is, for these techniques to work, the final result must be greater than or equal to zero. In the case of unsigned binary numbers the reason is clear; by definition, an unsigned binary number can only be used to represent a positive value, but subtracting a larger value from a smaller value results in a negative value. Thus, we need a way to represent negative numbers.

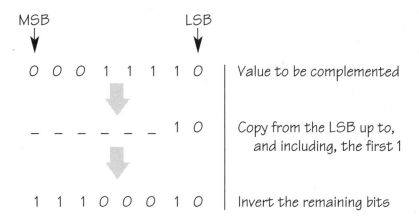

MSB   LSB

0 0 0 1 1 1 1 0 | Value to be complemented

– – – – – – 1 0 | Copy from the LSB up to,
                                        and including, the first 1

1 1 1 0 0 0 1 0 | Invert the remaining bits

Figure 8.7: Shortcut for generating a twos complement

Signed Binary Numbers

Signed binary numbers can be used to represent both positive and negative values, and they do this in a rather cunning way. In standard decimal arithmetic, negative numbers are typically represented in *sign-magnitude* form by prefixing the value with a minus sign; for example, a value of minus twenty-seven would be indicated as –27. However, for reasons of efficiency, computers rarely employ the sign-magnitude form, and instead use the signed binary format in which the most significant bit is also called the *sign bit* (Figure 8.8).

The least significant bits continue to represent the same positive quantities as for unsigned binary numbers, but the sign bit is used to represent a negative quantity. In the case of a signed 8-bit number, a 1 on the sign bit represents -2^7 (= –128), and the remaining bits are used to represent positive values in the range 0_{10} through $+127_{10}$. Thus, when the value represented by the sign bit is combined with the values represented by the remaining bits, an 8-bit signed binary number can be used to represent values in the range -128_{10} through $+127_{10}$.

To illustrate the differences between the sign-magnitude and signed binary formats, first consider a positive sign-magnitude decimal number and its negative equivalent; for example, +27 and –27. As we see, the digits are identical for both cases and only the sign changes.

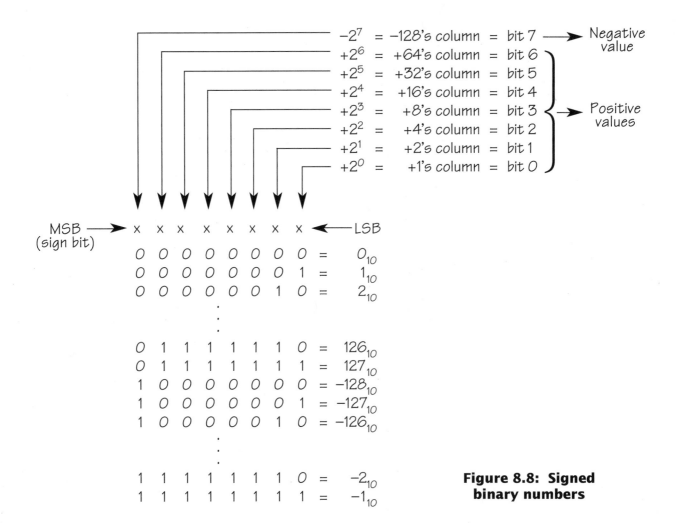

Figure 8.8: Signed binary numbers

Now consider the same values represented as signed binary numbers (Figure 8.9). In this case the bit patterns of the two binary numbers are very different. This is because the sign bit represents an actual quantity (-128_{10}) rather than a simple plus or minus; thus, the signed equivalent of -27_{10} is formed by combining -128_{10} with $+101_{10}$. Now pay attention because this is the clever part: closer investigation reveals that each bit pattern is in fact the twos complement of the other! To put this another way, taking the twos complement of a positive signed binary value returns its negative equivalent and vice versa.

The end result is that using signed binary numbers greatly reduces the complexity of operations within a computer. To illustrate why this is so, first

Figure 8.9: Comparison of positive and negative signed binary values

consider one of the simplest operations, that of addition. Compare the additions of positive and negative decimal values in sign-magnitude form with their signed binary counterparts (Figure 8.10). Examine the standard decimal calculations on the left. The one at the top is easy to understand because it's a straightforward addition of two positive values. However, even though you are familiar with decimal addition, you probably found the other three a little harder because you had to decide exactly what to do with the negative values. By comparison, the signed binary calculations on the right are all simple additions, regardless of whether the individual values are positive or negative.

Decimal sign-magnitude | Signed binary

	5	7		0 0 1 1 1 0 0 1
+	3	0		+ 0 0 0 1 1 1 1 0
=	8	7		0 1 0 1 0 1 1 1

	5	7		0 0 1 1 1 0 0 1
+	-3	0		+ 1 1 1 0 0 0 1 0
=	2	7		0 0 0 1 1 0 1 1

	-5	7		1 1 0 0 0 1 1 1
+	3	0		+ 0 0 0 1 1 1 1 0
=	-2	7		1 1 1 0 0 1 0 1

	-5	7		1 1 0 0 0 1 1 1
+	-3	0		+ 1 1 1 0 0 0 1 0
=	-8	7		1 0 1 0 1 0 0 1

Figure 8.10: Comparison of sign-magnitude versus signed binary additions

Similarly, if computers were forced to use a binary version of the sign-magnitude form to perform additions, they would have to perform a sequence of operations:

First compare the signs of the two values.	
IF THE SIGNS ARE THE SAME: Add the values. Note that the result will always have the same sign as the original values.	IF THE SIGNS ARE DIFFERENT: Subtract the smaller value from the larger value, then attach the correct sign to the result.

Figure 8.11: Comparison of sign-magnitude additions

As well as being time consuming, performing all these operations would require a substantial number of logic gates. Thus, the advantages of signed binary format for addition operations are apparent: signed binary numbers can always be directly added together to provide the correct result in a single operation, regardless of whether they represent positive or negative values. That is, the operations $a + b$, $a + -b$, $-a + b$, and $-a + -b$ are all performed in exactly the same way, by simply adding the two values together. This results in fast computers that can be constructed using a minimum number of logic gates.

Now consider the case of subtraction. We all know that $10 - 3 = 7$ in decimal arithmetic, and that the same result can be obtained by negating the right-hand value and inverting the operation: that is, $10 + -3 = 7$. This technique is also true for signed binary arithmetic, although the negation of the right hand value is performed by taking its twos complement rather than by changing its sign. For example, consider a generic[1] signed binary subtraction represented by $a - b$. Generating the twos complement of b results in $-b$, allowing the operation to be performed as an addition: $a + -b$.

---

[1] For the sake of simplicity, only the case of $a - b$ is discussed here. However, the operations $a - b$, $a - -b$, $-a - b$, and $-a - -b$ are all performed in exactly the same way, by simply taking the twos complement of b and adding the result to a, regardless of whether a and b represent positive or negative values.

This means that computers do not require different blocks of logic to add and subtract; instead, they only require an adder and a twos complementor. As well as being faster than a subtractor, a twos complementor requires significantly fewer logic gates.

Binary Multiplication

One technique for performing multiplication in any number base is by means of repeated addition; for example, in decimal, $6 \times 4 = 6 + 6 + 6 + 6 = 24$. However, even though computers can perform millions of operations every second, the repeated addition technique is time consuming when the values to be multiplied are large. As an alternative, binary numbers may by multiplied together by means of a shift-and-add technique known as *Booth's Algorithm*:

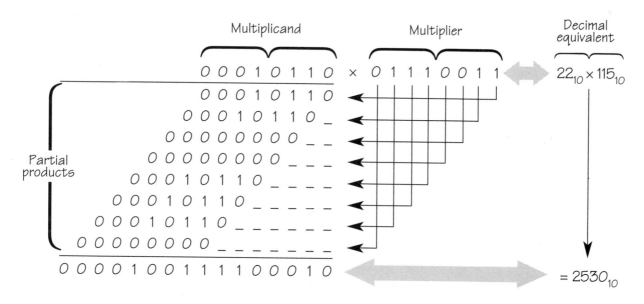

Figure 8.12: Binary multiplication using Booth's Algorithm

Using Booth's algorithm, a *partial product* is generated for every bit in the multiplier. If the value of the multiplier bit is 0, its corresponding partial product consists only of 0s; if the value of the bit is 1, its corresponding partial product is a copy of the multiplicand. Additionally, each partial product is left-shifted as a function of the multiplier bit with which it is associated; for example, the partial product associated with bit 0 in the multiplier is left-shifted zero bits, the partial product associated with bit 1 is left-shifted one bit, etc. All of the partial products are then added together to generate the result, whose width is equal to the sum of the widths of the two values being multiplied together.

There are several ways to implement a multiplier based on Booth's technique. In one implementation, all of the partial products are generated simultaneously and then added together. This requires a lot of logic gates, but the resulting multiplication is extremely fast. Unfortunately, Booth's Algorithm only works with unsigned binary values. However, this problem can be overcome by taking the twos complement of any negative values before feeding them into the multiplier. If the signs of the two values are the same, both positive or both negative, then no further action need be taken[2]—but if the signs are different, then the result returned from the multiplier must be negated by transforming *it* into its twos complement.

Finally, long division is approximately as much fun in binary as it is in decimal, which is to say, not a lot. For this reason, binary division is best left to computers, as they are in no position to argue about it.

---

[2] A negative multiplied by a negative equals a positive; for example, $-3 \times -4 = +12$.

Boolean Algebra

One of the most significant mathematical tools available to electronics designers was actually invented for quite a different purpose. Around the 1850s, a British mathematician, George Boole (1815–1864), developed a new form of mathematics that is now known as *Boolean Algebra*. Boole's intention was to use mathematical techniques to represent and rigorously test logical and philosophical arguments. His work was based on the following: a statement is a sentence that asserts or denies an attribute about an object or group of objects:

Statement: *Your face resembles a cabbage.*
Depending on how carefully you choose your friends, they may either agree or disagree with the sentiment expressed; therefore, this statement cannot be proved to be either true or false.

By comparison, a proposition is a statement that is either true or false with no ambiguity:

Proposition: *I just tipped a bucket of burning oil into your lap.*
This proposition may be true or it may be false, but it is definitely one or the other and there is no ambiguity about it.

Propositions can be combined together in several ways; propositions combined with an AND operator are known as a conjunction:

Conjunction: *You have a parrot on your head AND you have a fish in your ear.*

The result of a conjunction is true if all of the propositions comprising that conjunction are true.

Propositions combined with an OR operator are known as a disjunction:

Disjunction: *You have a parrot on your head OR you have a fish in your ear.*

The result of a disjunction is true if at least one of the propositions comprising that disjunction is true.

From these beginnings, Boole established a new mathematical field known as symbolic logic, in which the logical relationship between propositions can be represented symbolically by such means as equations or truth tables. Sadly, this work found little application outside the school of symbolic logic for almost one hundred years.

In fact, the significance of Boole's work was not fully appreciated until the late 1930s, when a graduate student at MIT, Claude Shannon, submitted a master's thesis that revolutionized electronics. In this thesis, Shannon showed that Boolean Algebra offered an ideal technique for representing the logical operation of digital systems. Shannon had realized that the Boolean concepts of FALSE and TRUE could be mapped onto the binary digits 0 and 1, and that both could be easily implemented by means of electronic circuits.

Logical functions can be represented using graphical symbols, equations, or truth tables, and these views can be used interchangeably (Figure 9.1).

There are a variety of ways to represent Boolean equations. In this book, the symbols &, | and ^ are used to represent AND, OR, and XOR respectively; a negation, or NOT, is represented by a horizontal line, or bar, over the portion of the equation to be negated.

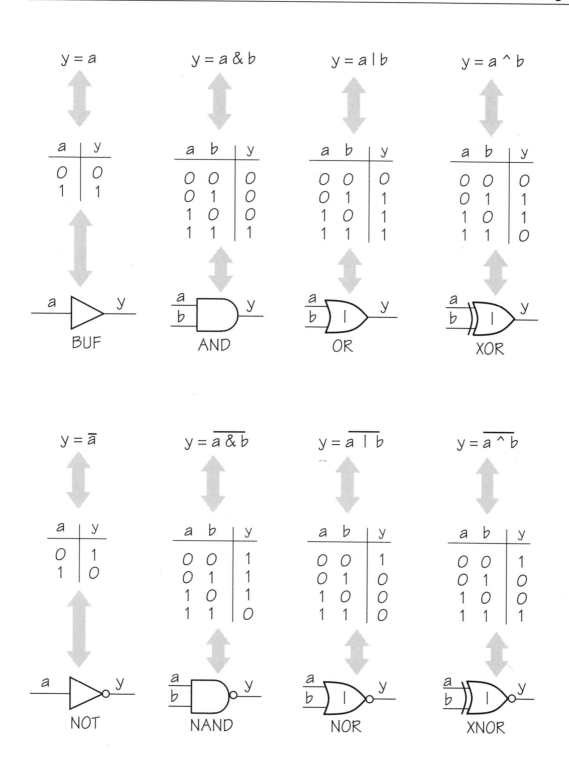

Figure 9.1: Summary of primitive logic functions

Combining a Single Variable with Logic 0 or Logic 1

A set of simple but highly useful rules can be derived from the combination of a single variable with a logic 0 or logic 1:[1,2]

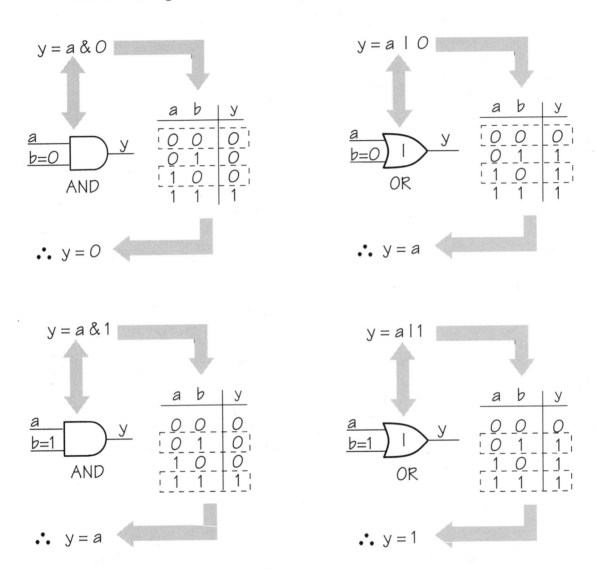

Figure 9.2: Combining a single variable with logic 0 or logic 1

---

[1] Note that, throughout these discussions, the results from the NAND and NOR functions would be the inverse of those from the AND and OR functions, respectively.

[2] Note that the symbol ∴ shown in the equations in Figure 9.2 means "*therefore*."

The Idempotent Rules

The rules derived from the combination of a single variable with itself are known as the idempotent rules:

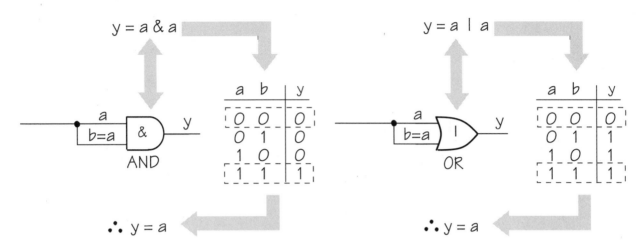

Figure 9.3: The idempotent rules

The Complementary Rules

The rules derived from the combination of a single variable with the inverse of itself are known as the complementary rules:

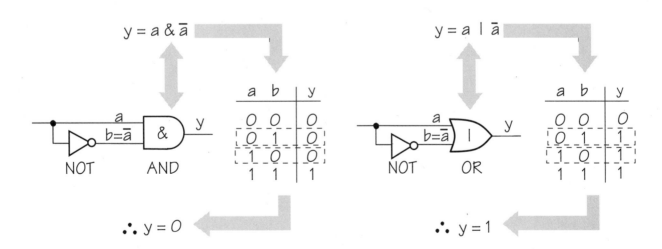

Figure 9.4: The complementary rules

The Involution Rule

The involution rule states that an even number of inversions cancel each other out; for example, two NOT functions connected in series generate an identical result to that of a BUF function:

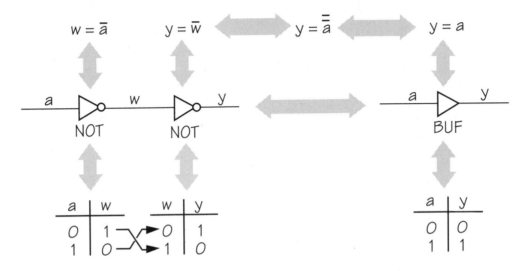

Figure 9.5: The involution rule

The Commutative Rules

The commutative rules state that the order in which variables are specified will not affect the result of an AND or OR operation:

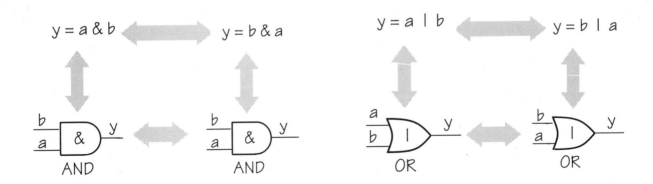

Figure 9.6: The commutative rules

The Associative Rules

The associative rules state that the order in which pairs of variables are associated together will not affect the result of multiple AND or OR operations:

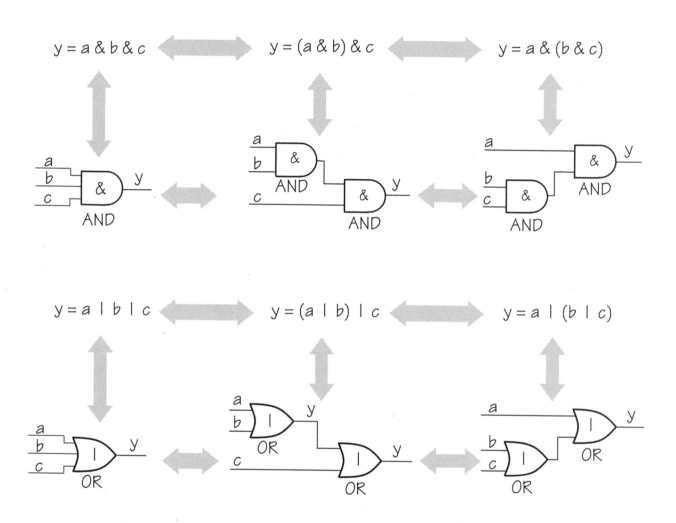

Figure 9.7: The associative rules

In addition to recognizing the application to Boolean algebra to electronic design, Shannon is also credited with the invention of the rocket-powered frisbee, and is famous for riding down the corridors at Bell Laboratories on a unicycle while simultaneously juggling four balls.

Precedence of Operators

In standard arithmetic, the multiplication operator is said to have a higher precedence than the addition operator. This means that, if an equation contains both multiplication and addition operators without parenthesis, then the multiplication is performed before the addition; for example:[3]

$$6 + 2 \times 4 \equiv 6 + (2 \times 4)$$

Similarly, in Boolean Algebra, the & operator has a higher precedence than the | operator:

$$a \mid b \& c \equiv a \mid (b \& c)$$

Due to the similarities between these arithmetic and logical operators, the & operator is known as a logical multiplication or *product*, and the | operator is known as a logical addition or *sum*. To avoid any confusion as to the order in which logical operations will be performed, this book will always make use of parentheses.

The first true electronic computer, ENIAC (Electronic Numerical Integrator and Calculator), was constructed at the University of Pennsylvania between 1943 and 1946. In many ways ENIAC was a monster; it occupied 30 feet by 50 feet of floor space, weighed approximately 30 tons, and used more than 18,000 vacuum tubes which required 150 kilowatts of power—enough to light a small town. One of the big problems with computers built from vacuum tubes was reliability; 90% of ENIAC's down-time was attributed to locating and replacing burnt-out tubes. Records from 1952 show that approximately 19,000 vacuum tubes had to be replaced in that year alone; that averages out to about 50 tubes a day!

---

[3] Note that the symbol ≡ shown in the equations indicates *"is equivalent to"* or *"is the same as."*

The First Distributive Rule

In standard arithmetic, the multiplication operator will distribute over the addition operator because it has a higher precedence; for example:

$$6 \times (5 + 2) \equiv (6 \times 5) + (6 \times 2)$$

Similarly, in Boolean Algebra, the & operator will distribute over an | operator because it has a higher precedence; this is known as the first distributive rule:

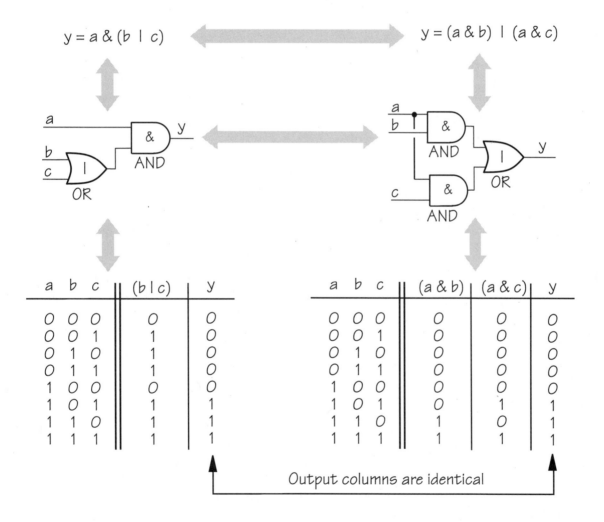

Figure 9.8: The first distributive rule

The Second Distributive Rule

In standard arithmetic, the addition operator will not distribute over the multiplication operator because it has a lower precedence:[4]

$$6 + (5 \times 2) \neq (6 + 5) \times (6 + 2)$$

However, Boolean Algebra is special in this case. Even though the | operator has lower precedence than the & operator, it will still distribute over the & operator; this is known as the second distributive rule:

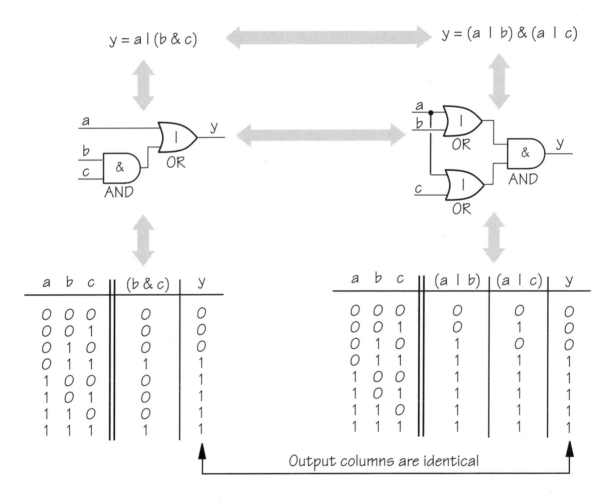

Figure 9.9: The second distributive rule

---

[4] Note that the symbol ≠ shown in the equation indicates *"is not equal to."*

The Simplification Rules

There are a number of simplification rules which can be used to reduce the complexity of Boolean expressions. As the end result is to reduce the number of logic gates required to implement the expression, the process of simplification is also known as *minimization:*

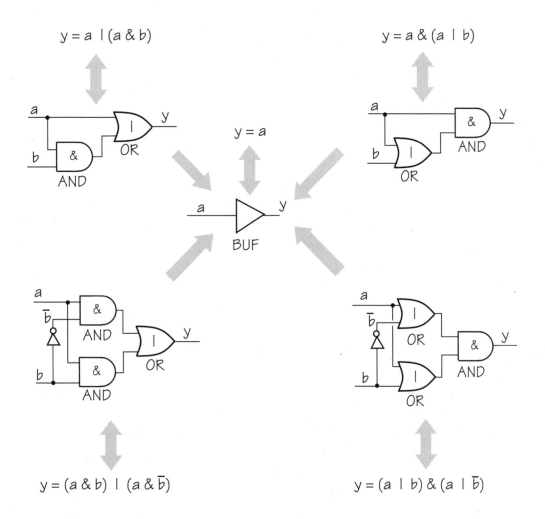

Figure 9.10: The simplification rules

– continued on next page

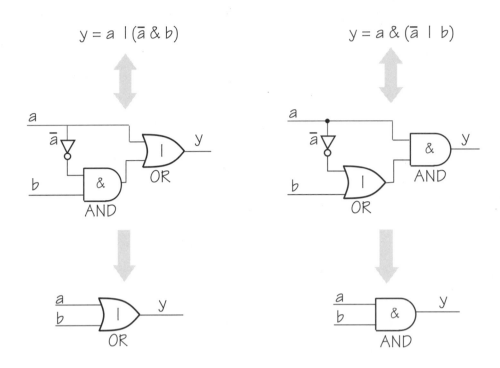

Figure 9.10 *(continued):* **The simplification rules**

Few now know that, before elbowing his way up the ladder of corporate success, the author of this epic volume once aspired to be a world-famous explorer. Picture him, if you will, as he strode across the veldt (hair by Vidal Sasson™, eyelashes by Max Factor™, and loin cloths by Fruit of the Loom™). The name Tracker Max, as he then styled himself, was oft bandied by the denizens of the dim and dingy taverns in the deepest, direst domains of the dark continent.

Today his Arnold Schwarzenegger look-alike body is compressed into an executive-style business suit (in a fashion reminiscent of years gone by) and he passes his twilight years quaffing the odd fermented coconut and penning trail-blazing books such as this one. Now that time has healed the deeper wounds, some in unusual and interesting places that he is only too happy to display (for a nominal fee), he feels more able to tell the tale of those far off times . . . unfortunately we don't have enough space here!

DeMorgan Transformations

A contemporary of Boole's, Augustus DeMorgan (1806–1871), also made significant contributions to the field of symbolic logic, most notably a set of rules which facilitate the conversion of Boolean expressions into alternate and often more convenient forms. A DeMorgan Transformation comprises four steps:

1. Exchange the & operators for | operators and vice versa.

2. Invert all the variables; also exchange 0s for 1s and vice versa.

3. Invert the entire function.

4. Reduce all multiple inversions.

Consider the DeMorgan Transformation of a 2-input AND function (Figure 9.11). Note that the NOT gate on the output of the new function can be combined with the OR to form a NOR.

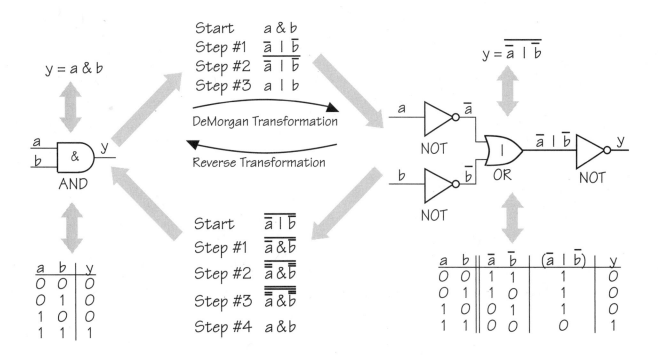

Figure 9.11: DeMorgan Transformation of an AND

Similar transformations can be performed on the other primitive functions:

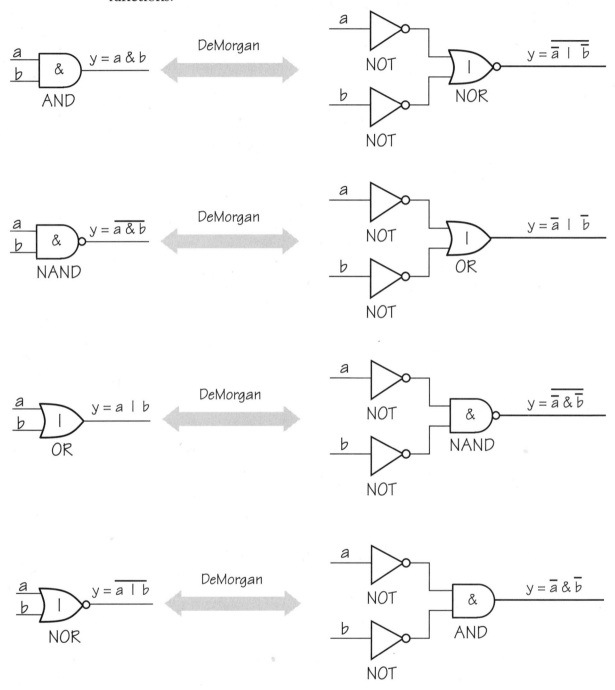

Figure 9.12: DeMorgan Transformations of AND, NAND, OR and NOR

Minterms and Maxterms

For each combination of inputs to a logical function, there is an associated *minterm* and an associated *maxterm*. Consider a truth table with three inputs a, b, and c (Figure 9.13.)

The minterm associated with each input combination is the &, or product, of the input variables; the maxterm is the |, or sum, of the inverted input variables. Minterms and maxterms are useful for deriving Boolean equations from truth tables as discussed below.

a	b	c	minterms	maxterms
0	0	0	$(\bar{a}\ \&\ \bar{b}\ \&\ \bar{c})$	$(a\ |\ b\ |\ c)$
0	0	1	$(\bar{a}\ \&\ \bar{b}\ \&\ c)$	$(a\ |\ b\ |\ \bar{c})$
0	1	0	$(\bar{a}\ \&\ b\ \&\ \bar{c})$	$(a\ |\ \bar{b}\ |\ c)$
0	1	1	$(\bar{a}\ \&\ b\ \&\ c)$	$(a\ |\ \bar{b}\ |\ \bar{c})$
1	0	0	$(a\ \&\ \bar{b}\ \&\ \bar{c})$	$(\bar{a}\ |\ b\ |\ c)$
1	0	1	$(a\ \&\ \bar{b}\ \&\ c)$	$(\bar{a}\ |\ b\ |\ \bar{c})$
1	1	0	$(a\ \&\ b\ \&\ \bar{c})$	$(\bar{a}\ |\ \bar{b}\ |\ c)$
1	1	1	$(a\ \&\ b\ \&\ c)$	$(\bar{a}\ |\ \bar{b}\ |\ \bar{c})$

Figure 9.13: Minterms and maxterms

Sum-of-Products and Product-of-Sums

A designer will often specify portions of a design using truth tables, and determine how to implement these functions as logic gates later. The designer may start by representing a function as a *black box* with an associated truth table. Note that the values assigned to output y in the truth table shown below were selected randomly, and have no significance beyond the purposes of this example.

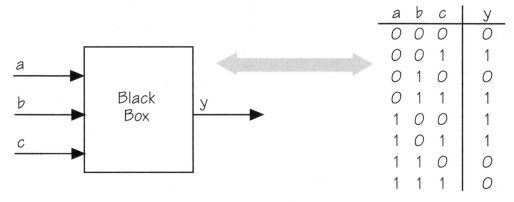

a	b	c	y
0	0	0	0
0	0	1	1
0	1	0	0
0	1	1	1
1	0	0	1
1	0	1	1
1	1	0	0
1	1	1	0

Figure 9.14: Black box with associated truth table

There are two commonly used techniques for deriving Boolean equations from a truth table. In the first technique, the minterms corresponding to each line in the truth table for which the output is a logic 1 are extracted and combined using | operators; this method results in an equation said to be in *sum-of-products* form. In the second technique, the maxterms corresponding to each line in the truth table for which the output is a logic 0 are combined using & operators; this method results in an equation said to be in *product-of-sums* form:

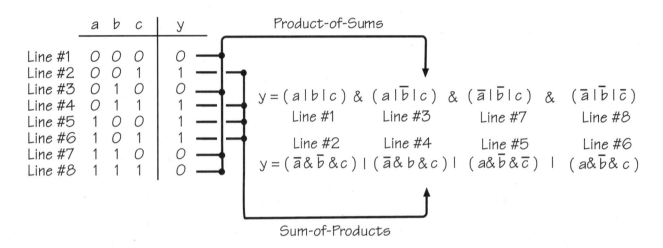

Figure 9.15: Sum-of-products versus product-of-sums forms

For a function whose output is logic 1 fewer times than it is logic 0, it is generally easier to extract a sum-of-products equation. Similarly, if the output is logic 0 fewer times than it is logic 1, it is generally easier to extract a product-of-sums equation. The sum-of-products and product-of-sums forms complement each other and return identical results. An equation in either form can be transformed into its alternative form by means of the appropriate DeMorgan Transformation.

Once an equation has been obtained in the required form, the designer would typically make use of the appropriate simplification rules to minimize the number of logic gates required to implement the function. However, neglecting any potential minimization, the equations above could be translated directly into their logic gate equivalents (Figure 9.16).

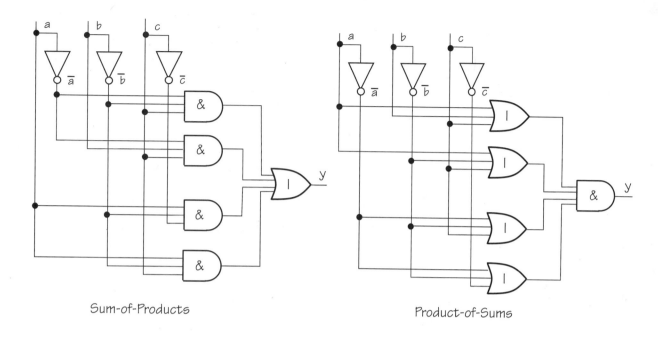

Sum-of-Products   Product-of-Sums

Figure 9.16: Sum-of-products versus product-of-sums implementations

Canonical Forms

In a mathematical context, the term *canonical form* is taken to mean a generic or basic representation. Canonical forms provide the means to compare two expressions without falling into the trap of trying to compare "apples" with "oranges." The sum-of-products and product-of-sums representations are both canonical forms. Thus, to compare two Boolean equations, both must first be coerced into the same canonical form; either sum-of-products or product-of-sums.

Karnaugh Maps

Karnaugh maps provide an alternative technique for representing Boolean functions; for example, consider the Karnaugh map for a 2-input AND:

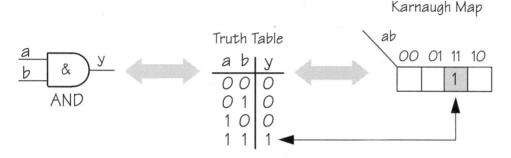

Figure 10.1: Karnaugh map for 2-input AND

The Karnaugh map comprises a box for every line in the truth table; the binary values above the boxes are those associated with the a and b inputs. Unlike a truth table, in which the input values typically follow a binary sequence: 00_2, 01_2, 10_2, and 11_2, the Karnaugh map's input values must be ordered so that the values for adjacent columns vary by only a single bit—for example, 00_2, 01_2, 11_2, and 10_2. This ordering is known as a *Gray code*, and it is a key factor in the way in which Karnaugh maps work. (Gray codes are discussed in more detail in Appendix D.)

The y column in the truth table shows all the 0 and 1 values associated with the gate's output. Similarly, all of the output values could be entered into the Karnaugh map's boxes. However, for reasons of clarity, it is common for only a single set of values to be used, typically the 1s.

Similar maps can be constructed for 3-input and 4-input functions. In the case of a 4-input map, the values associated with the c and d inputs must also be ordered as a Gray code; that is, they must be ordered in such a way that the values for adjacent rows vary by only a single bit (Figure 10.2).

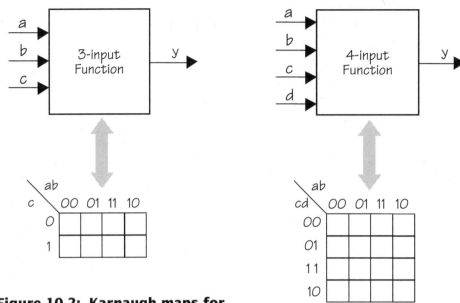

Figure 10.2: Karnaugh maps for 3-input and 4-input functions

Minimization Using Karnaugh Maps

Karnaugh maps often prove useful in the simplification and minimization of Boolean functions. Consider an example 3-input function represented as a black box with an associated truth table:[1]

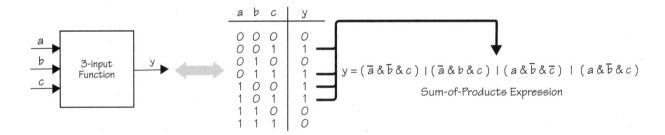

Figure 10.3: Example 3-input function

---

[1] The values assigned to output y in the truth table were selected randomly, and have no significance beyond the purposes of this example.

The equation extracted from the truth table in sum-of-products form contains four minterms, one for each of the 1s assigned to the output. Algebraic simplification techniques could be employed to minimize this equation, but this would necessitate every minterm being compared to each of the others, which can be somewhat time-consuming.

This is where Karnaugh maps enter the game. The 1s assigned to the map's boxes represent the same minterms as the 1s in the truth table's output column; however, as the input values associated with each row and column in the map differ by only one bit, any pair of horizontally or vertically adjacent boxes corresponds to minterms that differ by only a single variable. Such pairs of minterms can be grouped together and the variable that differs can be discarded:

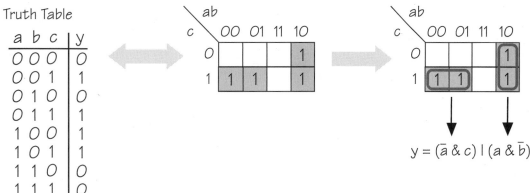

Figure 10.4: Karnaugh map minimization of example 3-input function

In the case of the horizontal group, input *a* is 0 for both boxes, input *c* is 1 for both boxes, and input *b* is 0 for one box and 1 for the other. Thus, for this group, changing the value on *b* does not affect the value of the output. This means that *b* is redundant and can be discarded from the equation representing this group. Similarly, in the case of the vertical group, input *a* is 1 for both boxes, input *b* is 0 for both boxes, and input *c* is 0 for one box and 1 for the other. Thus, for this group, input *c* is redundant and can be discarded.

Grouping Minterms

In the case of a 3-input Karnaugh map, any two horizontally or vertically adjacent minterms, each composed of three variables, can be combined to form a new product term composed of only two variables. Similarly, in the case of a 4-input map, any two adjacent minterms, each composed of four variables, can be combined to form a new product term composed of only three variables. Additionally, the 1s associated with the minterms can be used to form multiple groups. For example, consider a new 3-input function:

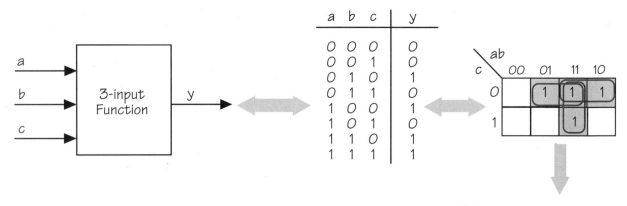

Figure 10.5: Karnaugh map minterms used to form multiple groups

$$y = (b \,\&\, \bar{c}) \mid (a \,\&\, \bar{c}) \mid (a \,\&\, b)$$

Groupings can also be formed from four adjacent minterms in which case two redundant variables can be discarded; consider some 4-input Karnaugh map examples (Figure 10.6).

In fact any group of 2^n adjacent minterms can be gathered together, where n is a positive integer. For example, 2^1 = two minterms, 2^2 = four minterms, 2^3 = eight minterms, etc.

As noted above, Karnaugh map input values are ordered so that the values associated with adjacent rows and columns differ by only a single bit. One result of this ordering is that the top and bottom rows are also only separated by a single bit; similarly, the left and right columns are only separated by a single bit. It may help to visualize the map rolled into a

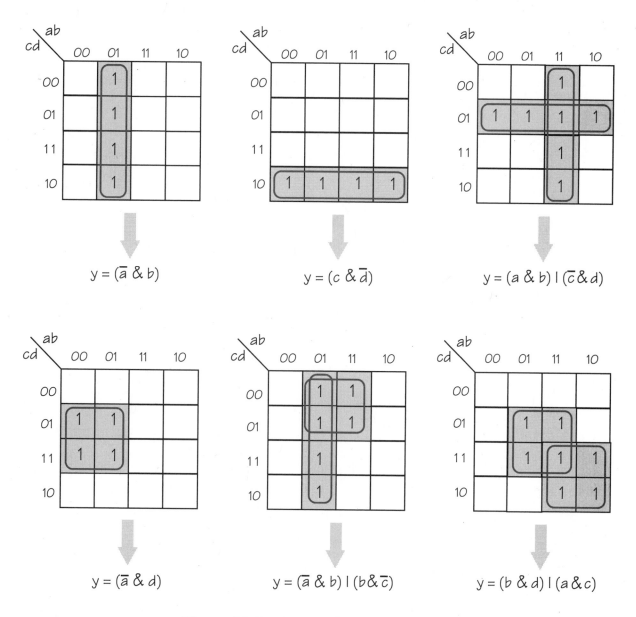

**Figure 10.6: Karnaugh map groupings
of four adjacent minterms**

horizontal cylinder such that the top and bottom edges are touching, or into
a vertical cylinder such that the left and right edges are touching. This leads
to some additional groupings, a few of which are shown in Figure 10.7.

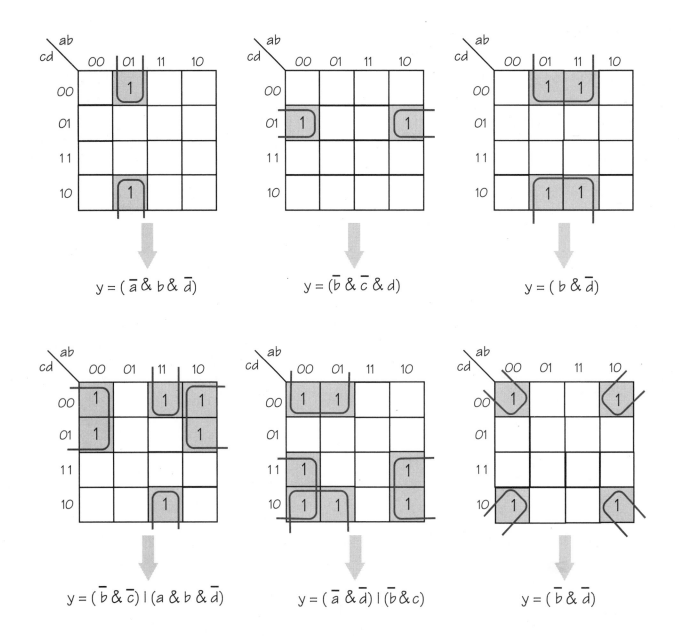

Figure 10.7: Additional Karnaugh map grouping possibilities

Note especially the last example. Diagonally adjacent minterms generally cannot be used to form a group, however, remembering that the left-right columns and the top-bottom rows are logically adjacent, the four corner minterms can be used to form a single group.

Incompletely Specified Functions

In certain cases a function may be incompletely specified; that is, the output may be undefined for some of the input combinations. If the designer knows that certain input combinations will never occur, then the value assigned to the output for these combinations is irrelevant. Alternatively, for some input combinations the designer may simply not care about the value on the output. In both cases, the designer can represent the output values associated with the relevant input combinations as question marks in the map (Figure 10.8).

The *?* characters indicate *don't care* states which can be considered to represent either 0 or 1 values at the designer's discretion. It should be noted that many electronics references use X characters to represent *don't care* states. Unfortunately, this may lead to confusion as design tools such as logic simulators use X characters to represent *don't know* states. Unless otherwise indicated, this book will use *?* and X to represent *don't care* and *don't know* states, respectively.

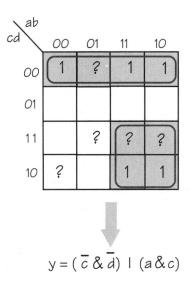

$$y = (\overline{c} \& \overline{d}) \mid (a \& c)$$

Figure 10.8: Karnaugh map for an incompletely specified function

Populating Maps Using 0s Versus 1s

When a Karnaugh map is populated using the 1s assigned to the truth table's output, the resulting Boolean expression is extracted from the map in sum-of-products form. As an alternative, the Karnaugh map can be populated using the 0s assigned to the truth table's output. In this case, groupings of 0s are used to generate expressions in product-of-sums format (Figure 10.9).

a	b	c	y
0	0	0	0
0	0	1	1
0	1	0	0
0	1	1	1
1	0	0	1
1	0	1	1
1	1	0	0
1	1	1	0

Sum-of-Products

$$y = (\overline{a} \& c) \mid (a \& \overline{b})$$

Equivalent

$$y = (a \mid c) \& (\overline{a} \mid \overline{b})$$

Product-of-Sums

Figure 10.9: Karnaugh maps populated using 0s versus 1s

Although the sum-of-products and product-of-sums expressions appear to be somewhat different, they do produce identical results. The expressions can be shown to be equivalent using algebraic means, or by constructing truth tables for each expression and comparing the outputs.

Karnaugh maps are most often used to represent 3-input and 4-input functions. It is possible to create similar maps for 5-input and 6-input functions, but these maps can become unwieldy and difficult to use. The Karnaugh technique is generally not considered to have any application for functions with more than six inputs.

Using Primitive Logic Functions to Build More Complex Functions

The primitive functions NOT, AND, OR, NAND, NOR, XOR, and XNOR can be connected together to build more complex functions which may, in turn, be used as building blocks in yet more sophisticated systems. The examples introduced below were selected because they occur commonly in designs, are relatively simple to understand, and are useful in later discussions.

Scalar versus Vector Notation

A single signal carrying one bit of binary data is known as a *scalar*. A set of signals carrying similar data can be gathered together into a group known as a *vector* (see the glossary definition of "vector" for more details):

Vector Notation
(Compressed)

Scalar Notation

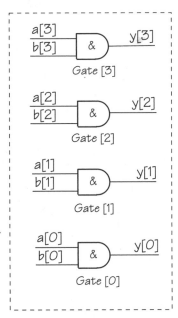

Vector Notation
(Expanded)

Figure 11.1: Scalar versus vector notation

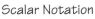

Each of the circuits above represent four 2-input AND gates. With scalar notation, each signal is assigned a unique name; for example, a3, a2, a1, and a0. Using vector notation, a single name is applied to a group of signals, and individual signals within the group are referenced by means of an index; for example, a[3], a[2], a[1], and a[0]. This means that a3 and a[3] represent two completely different signals; the former indicates a scalar named "a3," while the latter indicates the third element of a vector named "a."

Vector notation allows all the signals comprising the vector to be easily referenced in a single statement; for example, a[3:0], b[3:0], and y[3:0]. Thus, vector notation can be used to reduce the size and complexity of a circuit diagram while increasing its clarity.

Equality Comparators

In some designs it may be necessary to compare two sets of binary values to see if they contain the same data. Consider a function used to compare two 4-bit vectors a[3:0] and b[3:0]. A scalar output, *equal*, is to be set to logic 1 if each bit in a[3:0] is equal to the corresponding bit in b[3:0]; that is, the vectors are equal if a[3] = b[3], a[2] = b[2], a[1] = b[1], and a[0] = b[0] (Figure 11.2).

The values on a[3] and b[3] are compared using a 2-input XNOR gate. If the values are the same (both 0s or both 1s), the output of the XNOR will be 1; if the values are different, the output of the XNOR will be 0. Similar comparisons are performed between the other inputs, a[2] with b[2], a[1] with b[1], and a[0] with b[0]. The final AND gate is used to gather the results of the individual comparisons. If all the inputs to the AND gate are 1, the two vectors are the same and the output of the AND gate will be 1. Correspondingly, if any of the inputs to the AND gate are 0, the two vectors are different and the output of the AND gate will be 0.

A similar result could have been obtained by replacing the XNORs with XORs and the AND with a NOR. Either of these implementations could be easily extended to accommodate input vectors of greater width.

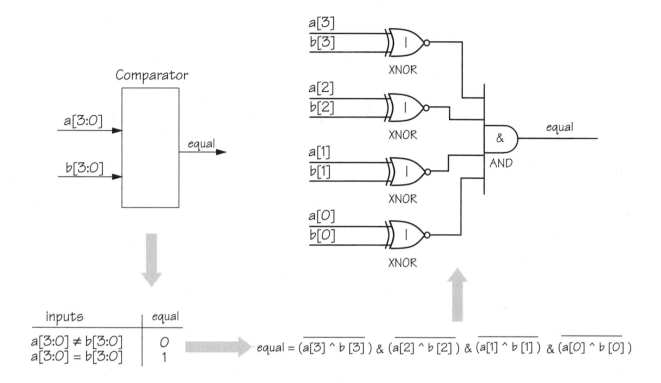

$$equal = \overline{(a[3] \wedge b\,[3]\,)} \ \& \ \overline{(a[2] \wedge b\,[2]\,)} \ \& \ \overline{(a[1] \wedge b\,[1]\,)} \ \& \ \overline{(a[0] \wedge b\,[0]\,)}$$

Figure 11.2: Equality comparator

Multiplexers

A multiplexer uses a binary value, or address, to select between a number of inputs and convey the data from the selected input to the output; for example, consider a 2:1 multiplexer (Figure 11.3). The *0* and *1* annotations on the multiplexer symbol represent the possible values of the *select* input and are used to indicate which data input will be selected.

The *?* characters in the truth table indicate *don't care* states. When *select* equals 0, the output from the function depends only on the value of *d0*, and we *don't care* about the value on *d1*. Similarly, when *select* equals 1, the output from the function depends only on the value of *d1*, and we *don't care* about the value on *d0*. The use of *don't care* states reduces the size of the truth table, better represents the operation of this particular function, and simplifies the extraction of the sum-of-products expression because the *don't cares* are ignored.

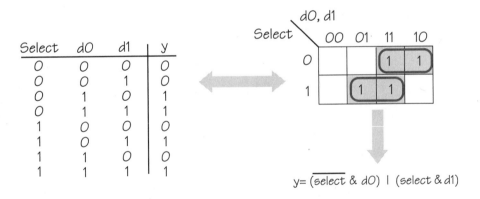

Figure 11.3: Multiplexer

An identical result could have been achieved using a full truth table combined with a Karnaugh map minimization:

Figure 11.4: Multiplexer equation extracted from a Karnaugh map

Larger multiplexers are also common in designs; for example, 4:1 multiplexers with two select inputs and four data inputs, 8:1 multiplexers with three select inputs and eight data inputs, etc.

Decoders

A decoder uses a binary value, or address, to select between a number of outputs and to assert the selected output by placing it in its active state; for example, consider a 2:4 decoder:

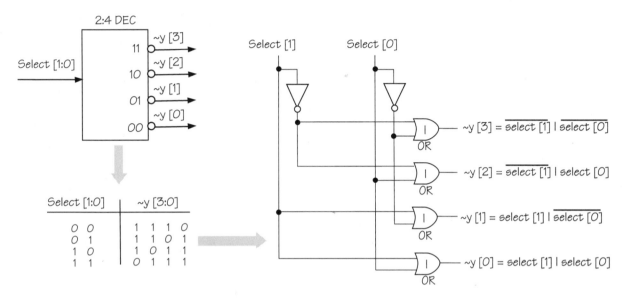

Figure 11.5: Decoder (active-low outputs)

The *00, 01, 10* and 11 annotations on the decoder symbol represent the possible values of the select[1:0] inputs and are used to indicate which output will be asserted.

The truth table shows that when a particular output is selected it is asserted to a 0, and when that output is not selected it returns to a 1. Because the outputs are asserted to 0s, this device is said to have *active-low* outputs. An active-low signal is one whose active state is considered to be a logic 0—similar functions can be created with *active-high* outputs. The active-low nature of this particular function is also indicated by the bobbles associated with the symbol's outputs and by the tilda characters in the names of the output signals.[1] Additionally, as each output is 0 for only

---

[1] The tilda '~' characters prefixing the output names ~y[3], ~y[2], ~y[1], and ~y[0] are used to indicate that these signals are active-low. The use of tilda characters is discussed in detail in Appendix A.

one input combination, it is simpler to extract the equations in product-of-sums format.

Larger decoders are also commonly used in designs: for example, 3:8 decoders with three select inputs and eight outputs, 4:16 decoders with four select inputs and sixteen outputs, etc.

Tri-State Functions

There is a special category of gates called tri-state functions whose outputs can adopt three states: 0, 1, and Z. Consider a tri-state buffer:

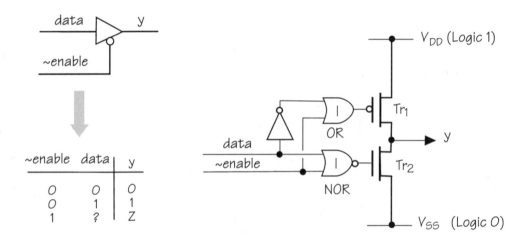

**Figure 11.6: Tri-state buffer
(active-low enable)**

The tri-state buffer's symbol is based on a standard buffer with an additional control input known as the enable. The active-low nature of this particular function's enable is indicated by the bobble associated with the symbol's input and by the tilda character in its name, ~enable. Similar functions with active-high enables may also be used in designs.

The Z character in the truth table represents a state known as *high-impedance*, in which the gate is not driving either of the standard 0 or 1 values. In fact, in the high-impedance state the gate is effectively disconnected from its output.

Although Boolean algebra is not well equipped to represent the Z state, the implementation of the tri-state buffer is relatively easy to understand. When the ~*enable* input is set to 1 (its inactive state), the output of the OR gate is forced to 1 and the output of the NOR gate is forced to 0, thereby turning both the Tr_1 and Tr_2 transistors OFF, respectively. With both transistors turned OFF, the output y is disconnected from V_{DD} and V_{SS}, and is therefore in the high-impedance state.

When ~*enable* is set to its active state of 0, the outputs of the OR and NOR gates are determined by the value on the *data* input. The circuit is arranged so that only one of the Tr_1 and Tr_2 transistors can be ON at any particular time. If *data* is set to 1, transistor Tr_1 is turned ON, thereby connecting output y to V_{DD}, or logic 1; if *data* is set to 0, transistor Tr_2 is turned ON, thereby connecting output y to V_{SS}, or logic 0.

Tri-state buffers can be used in conjunction with additional control logic to allow the outputs of multiple devices to drive a common signal (Figure 11.7). The use of the 2:4 decoder ensures that only one of the tri-state buffers is enabled at any time. The enabled buffer will propagate the data on its input to the common output, while the remaining buffers will be forced to their tri-state condition.

With hindsight it now becomes obvious that the standard primitive gates (AND, OR, NAND, NOR, etc.) depend on internal Z states to function; when any transistor is turned OFF, its output goes to a Z state. However, the standard primitive gates are constructed in such a way that at least one of the transistors connected to the output is turned ON; thus, the output of a standard gate is always driving either 0 or 1.

Figure 11.7: Using tri-state buffers

Combinational versus Sequential Functions

Logic functions are categorized as being either combinational (sometimes referred to as combinatorial) or sequential. In the case of a combinational function, the logic values on that function's outputs are directly related to the current combination of values on its inputs. All of the previous examples have been of this type.

In the case of a sequential function, the logic values on that function's outputs depend not only on its current input values, but also on previous input values; that is, the output values depend on a sequence of input values. Because sequential functions remember previous input values, they are also known as memory elements.

RS Latches

One of the simpler sequential functions is that of an RS latch, which can be implemented using two NOR gates connected in a back-to-back configuration:

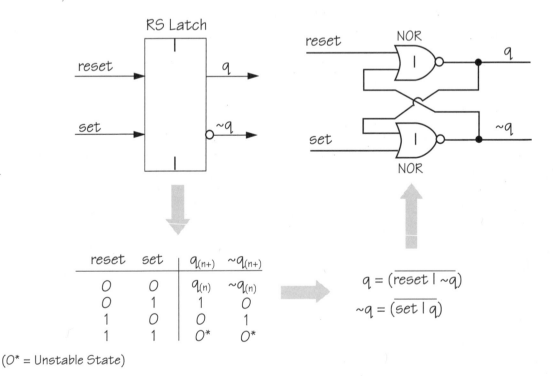

$$q = \overline{(reset \mid \sim q)}$$

$$\sim q = \overline{(set \mid q)}$$

(*0* = Unstable State*)

Figure 11.8: RS latch (NOR implementation)

In this NOR implementation, both *reset* and *set* inputs are active-high as indicated by the lack of bobbles on the symbol. The names of these inputs reflect the effect they have on the q output; when *reset* is active q is reset to 0, and when *set* is active q is set to 1. The q and ~q outputs are known as the *true* and *complementary* outputs, respectively.[2] In the latch's normal mode of operation, the value on ~q is usually the inverse, or complement, of the value on q; this is also indicated by the bobble associated with ~q on the symbol. The only time ~q is not the inverse of q occurs when both reset and set are active at the same time; this unstable state is discussed in more detail below.

The truth table labels $q_{(n+)}$ and $\sim q_{(n+)}$ indicate that these columns refer to the future values on the outputs. The n+ subscripts represent some future time, or *now-plus*. The labels $q_{(n)}$ and $\sim q_{(n)}$ indicate the current values on the outputs. The n subscripts represent the current time, or *now*. Thus, the first line of the truth table indicates that when both *reset* and *set* are inactive the future values on the outputs will be the same as their current values.

The secret of the RS latch's ability to remember previous input values is based on *feedback*—that is, the feeding back of the outputs as additional inputs into the function. Assume that both the *reset* and *set* inputs are initially in their inactive states, but that some previous input sequence placed the latch in its set condition—that is, q is 1 and ~q is 0. Now consider what occurs when the *reset* input goes to its active state and then returns to its inactive state (Figure 11.9).

As a reminder, if any input to a NOR is 1, the output is 0, and if both inputs to a NOR are 0, the output is 1. When *reset* goes to its active state ⎣ 1 ⎦, the q output from the first gate is forced to 0 ⎣ 2 ⎦. The 0 on q is fed back into the second gate ⎣ 3 ⎦ and, as both inputs to this gate are now 0, the ~q output is forced to 1 ⎣ 4 ⎦. The key point to note is that the 1 on ~q is now fed back into the first gate ⎣ 5 ⎦.

---

[2] In this case, the tilda '~' prefixing the output name ~q is used to indicate that this signal is a complementary output. Once again, the use of tilda characters is discussed in detail in Appendix A.

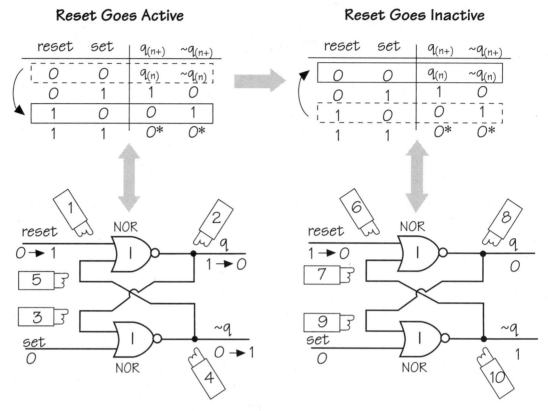

Reset Goes Active

reset	set	$q_{(n+)}$	$\sim q_{(n+)}$
0	0	$q_{(n)}$	$\sim q_{(n)}$
0	1	1	0
1	0	0	1
1	1	0*	0*

Reset Goes Inactive

reset	set	$q_{(n+)}$	$\sim q_{(n+)}$
0	0	$q_{(n)}$	$\sim q_{(n)}$
0	1	1	0
1	0	0	1
1	1	0*	0*

**Figure 11.9: RS latch
(reset goes active then inactive)**

When *reset* returns to its inactive state [6], the 1 from the $\sim q$ output continues feeding-back into the first gate [7] and the q output continues to be forced to 0 [8]. Similarly, the 0 on q continues feeding-back into the second gate [9] and the $\sim q$ output continues to be forced to 1 [10]. The end result is that the 1 at [7] causes the 0 at [8] which is fed back to [9], and the 0 on the *set* input combined with the 0 at [9] causes the 1 at [10] which is fed back to [7].

Thus, the latch has now been placed in its reset condition and a self-sustaining loop has been established. Even though both the *reset* and *set* inputs are now inactive, the q output remains at 0, indicating that *reset* was the last input to be in an active state. Once the function has been placed in its reset condition, any subsequent activity on the *reset* input will have no effect on the outputs and the only way to affect the function is by means of the *set* input.

Now consider what occurs when the *set* input goes to its active state and then returns to its inactive state (Figure 11.10). When *set* goes to its active state ⬚ 11 ✏, the ~q output from the second gate is forced to 0 ⬚ 12 ✏. The 0 on ~q is fed back into the first gate ⬚ 13 ✏ and, as both inputs to this gate are now 0, the q output is forced to 1 ⬚ 14 ✏. The key point to note is that the 1 on q is now fed back into the second gate ⬚ 15 ✏.

When *set* returns to its inactive state ⬚ 16 ✏, the 1 from the q output continues feeding-back to the second gate ⬚ 17 ✏ and the ~q output continues to be forced to 0 ⬚ 18 ✏. Similarly, the 0 on ~q continues feeding-back into the first gate ⬚ 19 ✏ and the q output continues to be forced to 1 ⬚ 20 ✏. The end result is that the 1 at ⬚ 17 ✏ causes the 0 at ⬚ 18 ✏ which is fed back to ⬚ 19 ✏, and the 0 on the *reset* input combined with the 0 at ⬚ 19 ✏ causes the 1 at ⬚ 20 ✏ which is fed back to ⬚ 17 ✏.

**Figure 11.10: RS latch
(set goes active then inactive)**

Thus, the latch has been returned to its set condition and, once again, a self-sustaining loop has been established. Even though both the reset and set inputs are now inactive, the q output remains at 1, indicating that set was the last input to be in an active state. Once the function has been placed in its set condition, any subsequent activity on the set input will have no effect on the outputs and the only way to affect the function is by means of the reset input.

The unstable condition indicated by the fourth line of the truth table occurs when both reset and set are active at the same time. Problems occur when both reset and set return to their inactive states simultaneously or too closely together (Figure 11.11).

When both reset and set are active at the same time, the 1 on reset ⌈ 21 ⌉ forces the q output to 0 ⌈ 22 ⌉ and the 1 on set ⌈ 23 ⌉ forces the ~q output to 0 ⌈ 24 ⌉. The 0 on q is fed back to the second gate ⌈ 25 ⌉ and the 0 on ~q is fed back to the first gate ⌈ 26 ⌉.

Now consider what occurs when reset and set go inactive simultaneously (⌈ 27 ⌉ and ⌈ 28 ⌉, respectively). When the new 0 values on reset and set are combined with the 0 values fed back from q ⌈ 29 ⌉ and ~q ⌈ 30 ⌉, each gate initially sees both of its inputs at 0 and therefore both gates attempt to drive their outputs to 1. After any delays associated with the gates have been satisfied, both the outputs will indeed go to 1. When the output of the first gate goes to 1, that value is fed back to the input of the second gate. While this is happening, the output of the second gate goes to 1, and that value is fed back to the input of the first gate. Each gate now has its fed-back input at 1, and both gates therefore attempt to drive their outputs to 0. The circuit has entered a *metastable* condition where the outputs oscillate between 0 and 1 values.

If both halves of the function were exactly the same, these metastable oscillations would continue indefinitely. But there will always be some differences between the gates and their delays, and the function will eventually collapse into either the reset condition or the set condition. As there is no way to predict the final values on the q and ~q outputs, they are indicated as being in X, or *don't know*, states (⌈ 29 ⌉ and ⌈ 30 ⌉).

**Figure 11.11: RS latch
(reset and set go inactive simultaneously)**

The X states will persist until a valid input sequence occurs on either the *reset* or *set* inputs.

An alternative implementation for an RS latch can be realized using two NAND gates connected in a back-to-back configuration (Figure 11.12). In a NAND implementation, both *~reset* and *~set* are active low as indicated by the bobbles on the symbol and by the tilda characters in their names. As a reminder, if any input to a NAND is 0, the output is 1, and if both inputs to a NAND are 1, the output is 0. You can work out the rest for yourself.

~ reset	~ set	$q_{(n+)}$	$\sim q_{(n+)}$
0	0	1*	1*
0	1	0	1
1	0	1	0
1	1	$q_{(n)}$	$\sim q_{(n)}$

(1* = Unstable State)

$$q = \overline{(\sim set \ \& \ \sim q)}$$

$$\sim q = \overline{(\sim reset \ \& \ q)}$$

**Figure 11.12: RS latch
(NAND implementation)**

D-Type Latches

A more sophisticated function called a D-type latch can be constructed by attaching two ANDs and a NOT to the front of an RS latch (Figure 11.13). The *enable* input is active high for this configuration, as indicated by the lack of a bobble on the symbol. When *enable* is set to 1 (its active state), the true and inverted versions of the *data* input are allowed to propagate through the AND gates and are presented to the back-to-back NOR gates. If the *data* input changes while *enable* is still active, the outputs will respond to reflect the new value.

When *enable* goes to 0 (its inactive state), it forces the outputs of both ANDs to 0, and any further changes on the *data* input have no effect. The back-to-back NOR gates remember the last value on *data* prior to *enable* going inactive. Consider an example waveform (Figure 11.14). While *enable* is active, the value on the *data* input appears on the *q* output and an inverted version appears on the *~q* output. As usual, there will always be some delay

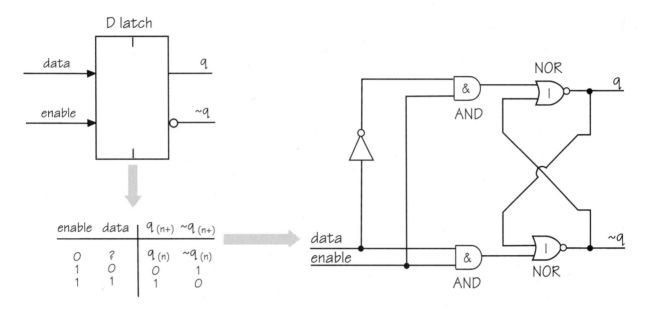

Figure 11.13: D-type latch (active-high enable)

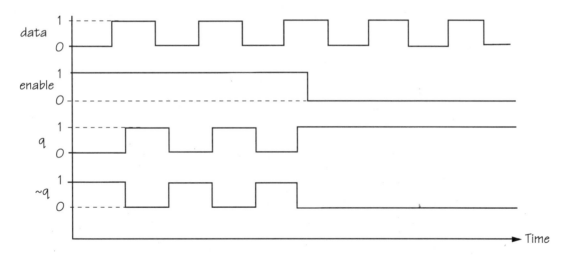

Figure 11.14: Waveform for D-type latch (active-high enable)

between changes on the inputs and corresponding responses on the outputs. When *enable* goes inactive, the outputs remember their previous values and no longer respond to any changes on the *data* input. As the operation of the device depends on the logic value, or level, on *enable*, this input is said to be *level-sensitive*.

D-Type Flip-flops

In the case of a D-type flip-flop, the data is loaded when a transition, or edge, occurs on the *clock* input. A transition from 0 to 1 is known as a *rising-edge* or a *positive-edge*, and a transition from 1 to 0 is known as a *falling-edge* or a *negative-edge*. D-type flip-flops may be positive-edge or negative-edge triggered (Figure 11.15).

The chevrons associated with the clock inputs on the symbols indicate that these are *edge-sensitive* inputs. A chevron without an associated bobble indicates a positive-edge clock, and a chevron with a bobble indicates a negative-edge clock. The last lines in the truth tables show that an inactive edge on the clock leaves the contents of the registers unchanged. These cases are often omitted from the truth tables.

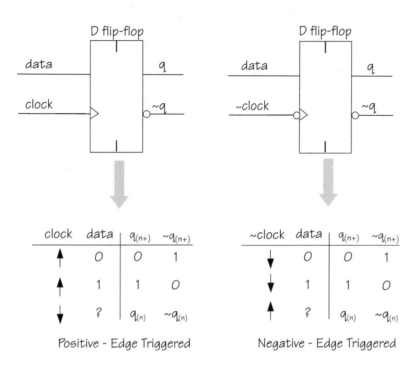

Figure 11.15: Positive- and negative-edge D-type flip-flops

Consider an example waveform for a positive-edge triggered D-type flip-flop (Figure 11.16). As the observer initially has no knowledge as to the contents of the register, the q and $\sim q$ outputs are initially shown as having X, or *don't know*, values. The first rising edge of the clock loads a 0 on the *data* input into the register; q goes to 0 and $\sim q$ goes to 1. The second rising edge of the clock loads a 1 on *data* into the register; q goes to 1 and $\sim q$ goes to 0.

Some registers have an additional input called $\sim clear$ or $\sim reset$ which forces q to 0 and $\sim q$ to 1, regardless of the value on *data*. The examples here illustrate active-low $\sim clear$ inputs, but active-high equivalents are also

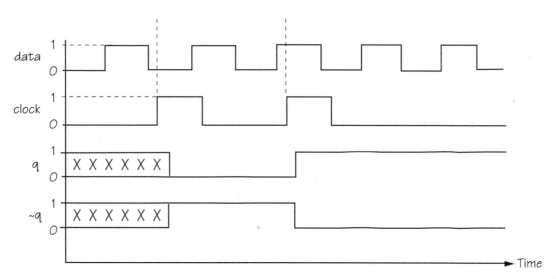

**Figure 11.16: Waveform for D-type flip-flop
(positive-edge triggered)**

available. Similarly, some registers have a ~*preset* or ~*set* input which forces q to 1 and ~q to 0, and some have both ~*clear* and ~*preset* inputs. These inputs may be either *asynchronous* or *synchronous* (Figure 11.17).

In the more common asynchronous case, the effect of ~*clear* going active is immediate and overrides both the *clock* and *data* inputs. In the synchronous case, the effect of ~*clear* is synchronized to the active edge of the *clock*.

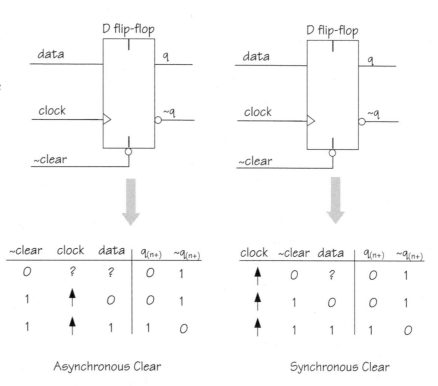

**Figure 11.17: D-type flip-flop
with clear**

JK and T Flip-flops

The majority of examples in this book are based on D-type flip-flops. However, for the sake of completeness, it should be noted that there are several other flavors of flip-flops available. Two common types are the JK and T, or Toggle, flip-flops:

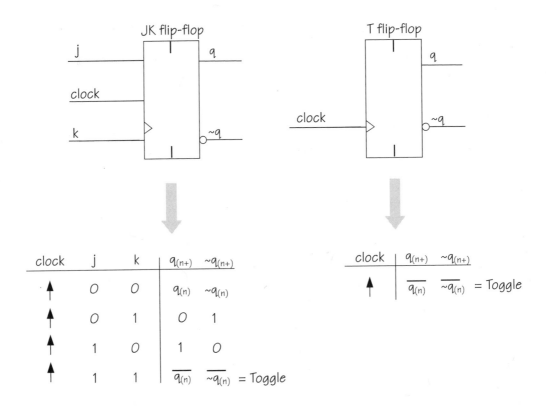

Figure 11.18: JK and T flip-flops

The first line of the JK flip-flop's truth table shows that, when both the j and k inputs are 0, an active edge on the *clock* leaves the contents of the register unchanged. The last line of the truth table shows that when both the j and k inputs are 1, an active edge on the *clock* causes the outputs to *toggle* to the inverse of their previous values. By comparison, the T flip-flop doesn't have any data inputs, the outputs simply toggle to the inverse of their previous values on each active clock edge.

Shift Registers

Shift registers allow binary data to be shifted one bit at a time and are commonly used in digital systems. Consider a simple 4-bit shift register constructed using D-type flip-flops:

Figure 11.19: SIPO shift register

This particular example is based on positive-edge triggered D-type flip-flops with active-low ~clear inputs. Also, this example is classed as a serial-in-parallel-out (SIPO) shift register, because data is loaded in serially and read out in parallel. When the ~clear input is set to 1 (its inactive state), a positive-edge on the clock input loads the value on serial_in into the first register, dff[0]. At the same time, the data that used to be in dff[0] is loaded into dff[1], the data that used to be in dff[1] is loaded into dff[2], and the data that used to be in dff[2] is loaded into dff[3].

For each flip-flop there is a delay between the active edge on the clock and the response at the output. These delays provide sufficient time for the next flip-flop in the chain to load the data from the previous stage before it changes. Consider an example waveform where a single logic 1 value is migrated through the shift register (Figure 11.20).

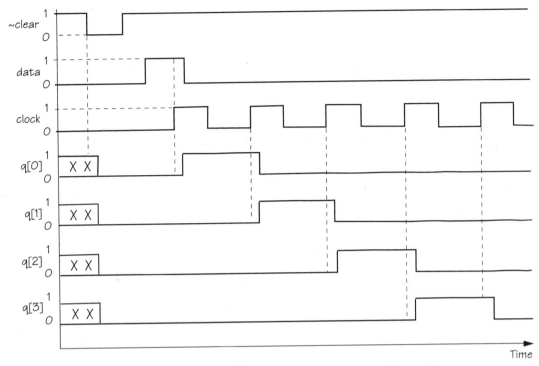

Figure 11.20: Waveform for SIPO shift register

Other common shift register variants are the parallel-in-serial-out (PISO), and the serial-in-serial-out (SISO); for example, consider a 4-bit SISO shift register:

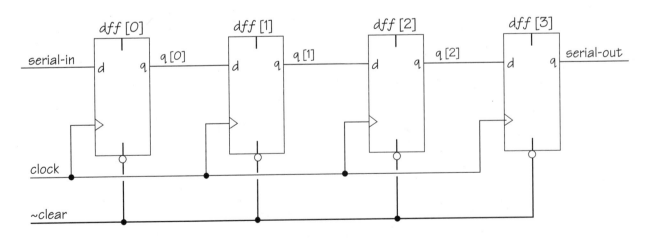

Figure 11.21: SISO shift register

Counters

Counters are also commonly used in digital systems. The number of states that the counter will sequence through before returning to its original value is called the *modulus* of the counter. For example, a function that counts from 0000_2 to 1111_2 in binary (or 0 to 15 in decimal) has a modulus of sixteen and would be called a modulo-16, or mod-16, counter. Consider a modulo-16 counter implemented using D-type flip-flops:

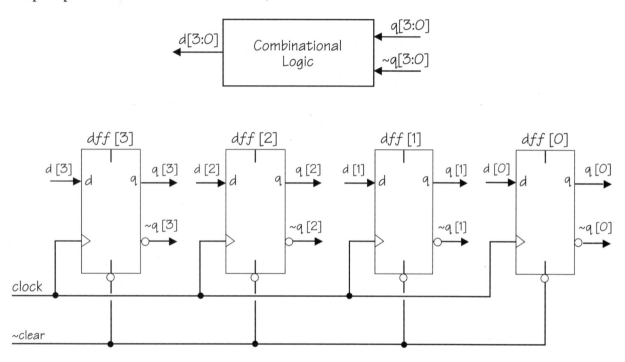

Figure 11.22: Binary counter

This particular example is based on positive-edge triggered D-type flip-flops with active-low ~clear inputs. The four flip-flops are used to store the current count value which is displayed on the q[3:0] outputs. When the ~clear input is set to 1 (its inactive state), a positive-edge on the clock input causes the counter to load the next value in the count sequence.

A block of combinational logic is used to generate the next value, d[3:0], based on the current value q[3:0] (Figure 11.23). Note that there is no need to create the inverted versions of q[3:0], because these signals are already available from the flip-flops as ~q[3:0].

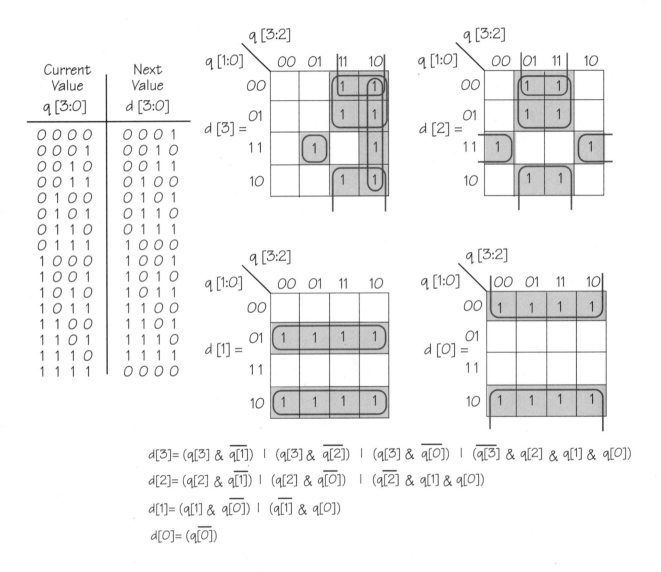

Figure 11.23: Generating the next count value

Brick by Brick

Let us pause here briefly for a philosophical moment. Consider a brick made of clay. Now, there's not a lot you can do with a single brick, but when you combine thousands of bricks together you can create the most tremendous structures. At the end of the day, the Great Wall of China is no more than a pile of bricks molded by man's imagination.[3]

In the world of the electronics engineer, transistors are the clay, primitive logic gates are the bricks, and the functions described above are simply building blocks. (We might also note that clay and transistors share something else in common—they both consist predominantly of silicon!) Any digital system, even one as complex as a supercomputer, is constructed from building blocks like comparators, multiplexers, shift registers, and counters. Once you understand the building blocks, there are no end to the things you can achieve!

---

[3] Approximately 2,400 km in length, the Great Wall is said to be the only man-made structure visible from space.

State Diagrams, State Tables, and State Machines

Consider a coin-operated machine that accepts nickels and dimes[1] and, for the princely sum of fifteen cents, dispenses some useful article called a gizmo that the well-dressed man-about-town could not possibly be without:

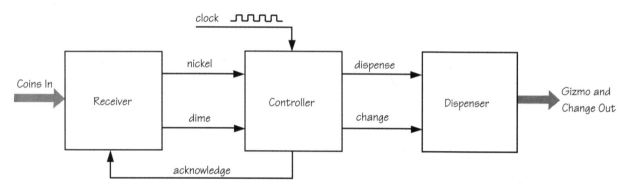

Figure 12.1: Block diagram of a coin-operated machine

The connections marked nickel, dime, dispense, change, and acknowledge represent digital signals carrying logic 0 and logic 1 values. The user can deposit nickels and dimes into the receiver in any order, but may only deposit one coin at a time. When a coin is deposited the receiver determines its type and sets the corresponding signal (nickel or dime) to a logic 1.

The operation of the controller is synchronized by the clock signal. On a rising edge of the clock, the controller examines the nickel and dime inputs to see if any coins have been deposited. The controller keeps track of the amount of money deposited and determines if any actions are to be performed.

---

[1] For the sake of those readers who do not reside in the United States, nickels and dimes are American coins worth five and ten cents, respectively.

Every time the controller inspects the *nickel* and *dime* signals, it sends an *acknowledge* signal back to the receiver. The *acknowledge* signal informs the receiver that the coin has been accounted for and the receiver responds by resetting the *nickel* and *dime* signals to 0 and awaiting the next coin. The *acknowledge* signal can be generated in a variety of ways which are not particularly relevant here.

When the controller decides that sufficient funds have been deposited, it instructs the dispenser to dispense a gizmo and any necessary change by setting the *dispense* and *change* signals to 1, respectively.

State Diagrams

A useful level of abstraction for a function such as the controller is to consider it as consisting of a set of *states* through which it sequences. The current state depends on the previous state combined with the previous values on the *nickel* and *dime* inputs. Similarly, the next state depends on the current state combined with the current values on the *nickel* and *dime* inputs. The operation of the controller may be represented by means of a *state diagram*, which offers a way to view the problem and to describe a solution (Figure 12.2).

The states are represented by the circles labeled 0-cents, 5-cents, 10-cents, 15-cents, and 20-cents, and the values on the *dispense* and *change* outputs are associated with these states. The arcs connecting the states are called *state transitions* and the values of the *nickel* and *dime* inputs associated with the state transitions are called *guard conditions*. The controller will only sequence between two states if the values on the *nickel* and *dime* inputs are the same as the guard conditions.

Assume that the controller is in its initial state of 0-cents. The values of the *nickel* and *dime* inputs are tested on every rising edge on the *clock*.[2] As long as no coins are deposited, the *nickel* and *dime* inputs remain at 0 and the controller remains in the 0-cents state. Once a coin is deposited, the next rising edge on the *clock* will cause the controller to sequence to the

---

[2] The controller is known to sequence between states only on the rising edge of the *clock*, so displaying this signal on every state transition would be redundant.

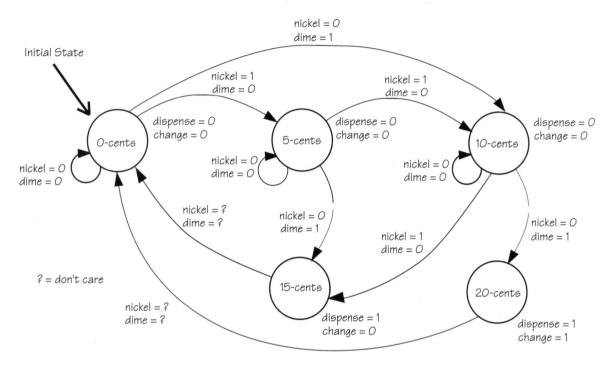

Figure 12.2: State diagram for the controller

5-cents or the 10-cents states depending on the coin's type. It is at this point that the controller sends an *acknowledge* signal back to the receiver instructing it to reset the *nickel* and *dime* signals back to 0 and to await the next coin.

State Tables

Another form of representation is that of a *state table*. This is similar to a truth table, but it also includes the current state as an input and the next state as an output (Figure 12.3).

In this instance the *clock* signal has been included for purposes of clarity. However, as for the state diagram, displaying this signal is somewhat redundant and it is usually omitted.

Current state	clock	nickel	dime	dispense	change	Next state
0-cents	↑	0	0	0	0	0-cents
0-cents	↑	1	0			5-cents
0-cents	↑	0	1			10-cents
5-cents	↑	0	0	0	0	5-cents
5-cents	↑	1	0			10-cents
5-cents	↑	0	1			15-cents
10-cents	↑	0	0	0	0	10-cents
10-cents	↑	1	0			15-cents
10-cents	↑	0	1			20-cents
15-cents	↑	?	?	1	0	0-cents
20-cents	↑	?	?	1	1	0-cents

Figure 12.3: State table for the controller

State Machines

The actual implementation of a function such as the controller is called a *state machine*. The heart of a state machine consists of a set of registers known as the *state variables*. Each state, 0-cents, 5-cents, 10-cents, ... is assigned a unique binary pattern of 0s and 1s, and the pattern representing the current state is stored in the state variables.

The two most common forms of synchronous, or clocked, state machines are known as *Moore* and *Mealy* machines after the men who formalized them. A Moore machine is distinguished by the fact that the outputs are derived only from the values in the state variables (Figure 12.4). The controller function featured in this discussion is a classic example of a Moore machine.

By comparison, the outputs from a Mealy machine may be derived from a combination of the values in the state variables and one or more of the inputs (Figure 12.5).

In both of the Moore and Mealy forms, the input logic consists of primitive gates such as AND, NAND, OR, and NOR. These combine the values on the inputs with the current state fed back from the state variables to generate the pattern of 0s and 1s representing the next state. This pattern of 0s and 1s is presented to the inputs of the state variables and is loaded on the next rising edge of the *clock*.

The output logic also consists of standard primitive logic gates that generate the appropriate values on the outputs from the current state stored in the state variables.

State Assignment

A key consideration in the design of a state machine is that of *state assignment*, the process by which the states are assigned to the binary patterns of 0s and 1s that are to be stored in the state variables.

A common form of state assignment requiring the minimum number of registers is *binary encoding*. A single register can only be assigned the two binary values of 0 or 1, two registers can be assigned four binary values, three registers can be assigned eight binary values, etc. The controller

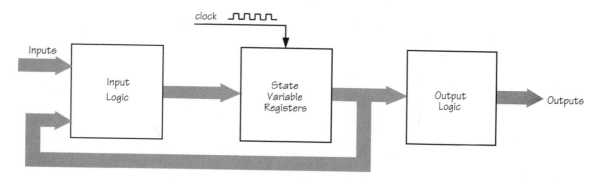

Figure 12.4: Block diagram of a Moore machine

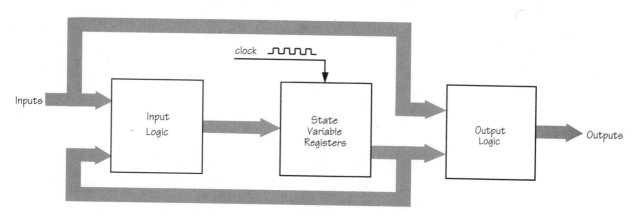

Figure 12.5: Block diagram of a Mealy machine

consists of five unique states, and therefore requires a minimum of three state variable registers.

The actual process of binary encoded state assignment is a nontrivial problem. In the case of the controller function, there are 6720 possible combinations[3] in which the five states can be assigned to the eight binary values provided by three registers. Each of these solutions may require a different arrangement of primitive gates to construct the input and output logic. Additionally, the type of registers used to implement the state variables also affects the supporting logic; the following discussions assume the use of D-type flip-flops.

---

[3] This number would be somewhat reduced if all the mirror-image combinations were taken into account, but this would not significantly affect the problem of determining the optimal combination.

Assuming that full use is made of *don't care* states, an analysis of the various binary encoded solutions for the controller yields the following:

138 solutions requiring 7 product terms
852 solutions requiring 8 product terms
1,876 solutions requiring 9 product terms
3,094 solutions requiring 10 product terms
570 solutions requiring 11 product terms
190 solutions requiring 12 product terms

where a product term is a group of literals linked by & operators, for example, (a & b & c), and a literal is any true or inverted variable. Thus, the product term (a & b & c) contains three literals.

A further analysis of the 138 solutions requiring seven product terms yields the following:

66 solutions requiring 17 literals
24 solutions requiring 18 literals
48 solutions requiring 19 literals

Thus, the chances of a random assignment resulting in an optimal solution is relatively slight. Fortunately, there are computer programs available to aid designers in this task. One solution resulting in the minimum number of product terms and literals is as follows:

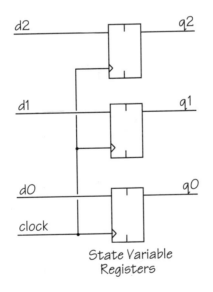

Current State			State Assignments
q2	q1	q0	
0	0	0	10-cents
0	0	1	15-cents
0	1	0	—
0	1	1	20-cents
1	0	0	0-cents
1	0	1	5-cents
1	1	0	—
1	1	1	—
d2	d1	d0	State Assignments
	Next State		

Figure 12.6: Binary encoded state assignment

A truth table for the controller function could now be derived from the state table (shown in Figure 12.3) by replacing the assignments in the *current state* column with the corresponding binary patterns for the state variable outputs ($q2$, $q1$, and $q0$), and replacing the assignments in the *next state* column with the corresponding binary patterns for the state variable inputs ($d2$, $d1$, and $d0$). The resulting equations could then be derived from the truth table by means of standard algebraic or Karnaugh map techniques. As an alternative, a computer program can be used to obtain the same results in less time with far fewer opportunities for error. Whichever technique is employed, the state assignments above lead to the following minimized Boolean equations:

$$d0 \ = \ \mathbf{(\overline{q0} \ \& \ \overline{q2} \ \& \ dime)} \ | \ (q0 \ \& \ q2 \ \& \ \overline{nickel}) \ | \ (\overline{q0} \ \& \ nickel)$$

$$d1 \ = \ \mathbf{(\overline{q0} \ \& \ \overline{q2} \ \& \ dime)}$$

$$d2 \ = \ \mathbf{(q0 \ \& \ \overline{q2})} \ | \ (q2 \ \& \ \overline{nickel} \ \& \ \overline{dime}) \ | \ (\overline{q0} \ \& \ q2 \ \& \ \overline{dime})$$

$$dispense \ = \ \mathbf{(q0 \ \& \ \overline{q2})}$$

$$change \ = \ (q1)$$

The product terms shown in bold appear in multiple equations; however, regardless of the number of times a product term appears, it is only counted once because it only has to be physically implemented once. Similarly, the literals used to form product terms which appear in multiple equations are only counted once.

Another common form of state assignment is known as *one-hot encoding*, in which each state is represented by an individual register; thus, our controller with its five states would require five register bits. The one-hot technique typically requires a greater number of logic gates than does binary encoding. But, as the logic gates are used to implement simpler equations, the one-hot method results in faster state machines that can operate at higher clock frequencies.

Don't Care States, Unused States, and Latch-Up Conditions

It was previously noted that the analysis of the binary encoded state assignment made full use of *don't care* states; this must now be discussed in more detail. Our definition of the coin-operated machine noted that only a single coin can be deposited at a time, which means that the *nickel* and *dime* signals will never be assigned 1 values simultaneously. The designer or computer program can use this information to assign *don't care* states to the outputs for any combination of inputs that includes a 1 on both *nickel* and *dime* signals.

Additionally, the three binary encoded state variable registers provide eight possible combinations, of which only five were used. The analysis above was based on the assumption that *don't care* states can be assigned to the outputs for any combination of inputs that includes one of the unused patterns on the state variables. This assumption also requires further justification.

When the coin-operated machine is first powered-up, each state variable register can potentially initialize with a random logic 0 or logic 1 value. The controller could therefore power-up with its state variables containing any of eight possible patterns of 0s and 1s. For some state machines this would not be an important consideration, but this is not true in the case of the coin-operated machine controller. For example, the controller could power-up in the state 20-cents, in which case it would immediately dispense a gizmo and five cents change. The owner of such a machine may well be of the opinion that this was a less than ideal feature. Alternatively, the controller could power-up with its state variables in one of the unused combinations. On subsequent clock edges the controller could sequence directly or via one or more of the other unused combinations to any of the defined states. In a worst-case scenario, the controller could remain in the unused combination indefinitely or sequence endlessly between unused combinations; these worst-case scenarios are known as *latch-up* conditions.

One method of avoiding latch-up conditions is to assign additional, dummy states to each of the unused combinations and to define state transitions from each of these dummy states to the controller's initialization state, 0-cents. Unfortunately, in the case of our coin-operated machine, this technique would not affect the fact that the controller could wake up in a valid state other than 0-cents. An alternative is to provide some additional circuitry to generate a *power-on-reset* signal—for example, a single pulse that occurs only when the power is first applied to the machine. The power-on-reset can be used to force the state variable registers into the pattern associated with the 0-cents state. The analysis above assumed the use of a power-on-reset.

Analog-to-Digital and Digital-to-Analog

As we began our discussions in Chapter 1 by separating the analog and digital views of the world, it seems appropriate to close this section by reuniting them. While some systems operate solely on digital data, others have to interact with the analog world. It may be necessary to convert an analog input into a form that can be manipulated by the digital system, or to transform an output from the digital system into the analog realm. These tasks are performed by *analog-to-digital (A/D)* and *digital-to-analog (D/A)* converters, respectively:

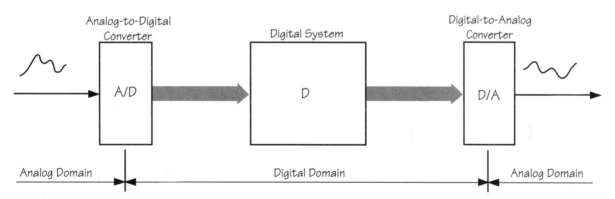

Figure 13.1: Analog-to-digital and digital-to-analog converters

Analog-to-Digital

A *transducer* is a device that converts input energy of one form into output energy of another. Analog effects can manifest themselves in a variety of different ways such as heat and pressure. To be processed by a digital system, the analog quantity must be detected and converted into a suitable form by means of an appropriate transducer called a *sensor*. For example, a microphone is a sensor which detects sound and converts it into a corresponding voltage. The analog-to-digital conversion process can be represented as follows:

Figure 13.2: Analog-to-digital conversion process

The output from the sensor typically undergoes some form of signal processing such as filtering and amplification before being passed to the A/D converter. This signal processing is generically known as *conditioning*. The A/D converter accepts the conditioned analog voltage and converts it into a series of equivalent digital values by *sampling* and *quantization* (Figure 13.3).

The sampling usually occurs at regular time intervals and is triggered by the digital part of the system. The complete range of values that the analog signal can assume is divided into a set of discrete bands or quanta. At each sample time the A/D converter determines which band the analog signal falls into (quantization) and outputs the equivalent binary code for that band. The main factor governing the accuracy of the conversion is the number of bands used; for example, a 3-bit code can only represent eight bands, each encompassing 12.5% of the analog signal's range, while a 12-bit code can represent 4096 bands, each encompassing 0.025% of the signal's range.

Digital-to-Analog

A D/A converter accepts a digital code and transforms it into a corresponding analog current or voltage suitable for use by an appropriate transducer called an *actuator*. For example, a loudspeaker is an actuator which converts an electrical signal into sound. The digital-to-analog conversion process can be represented as shown in Figure 13.4.

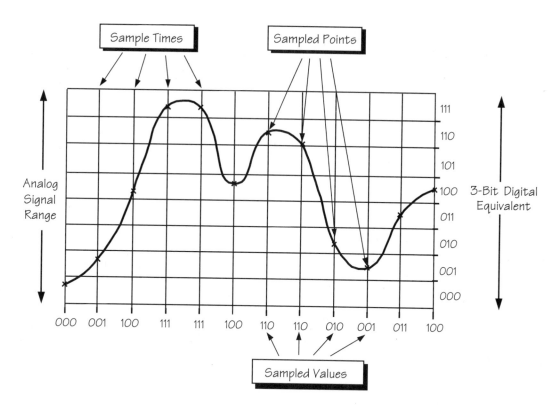

Figure 13.3: Sampling and quantizing an analog signal

The conversions usually occur at regular time intervals and are triggered by a clock signal from the digital part of the system. The output from the D/A converter typically undergoes some form of conditioning before being passed to the actuator.

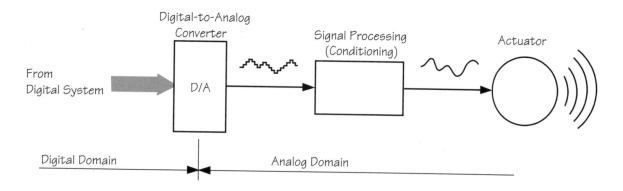

Figure 13.4: Digital-to-analog conversion process

Chapter 14

Integrated Circuits (ICs)

In the 1950s, transistors and other electronic components were only available in individual packages. These *discrete* components were laid out on a circuit board and hand connected using separate wires. At that time an electronic gate capable of storing a single binary bit of data cost more than $2. By comparison, in the early 1990s, enough gates to store 5000 bits of data cost less than a cent. This vast reduction in price was primarily due to the invention of the integrated circuit (IC).

A functional electronic circuit requires transistors, resistors, capacitors, diodes, and the connections between them. A monolithic integrated circuit has all these components formed on the surface layer of a sliver, or chip, of a single piece of semiconductor; hence the term monolithic, meaning seamless. Although a variety of semiconductor materials are available, the most commonly used is silicon and integrated circuits are popularly known as *silicon chips*. Unless otherwise noted, the remainder of these discussions will assume silicon as the semiconductor.

The Fabrication Process

The construction of integrated circuits requires one of most exacting production processes ever developed. The environment must be at least a thousand times cleaner than that of an operating theater, and impurities in materials have to be so low as to be measured in parts per billion. The process begins with the growing of a single crystal of pure silicon in the form of a cylinder with a diameter between 25 mm and 200 mm. The cylinder is cut into paper-thin slices called *wafers*, which are approximately 0.2 mm thick (Figure 14.1).

Figure 14.1: Creating silicon wafers

The thickness of the wafer is determined by the requirement for sufficient mechanical strength to allow it to be handled without damage. The actual thickness necessary for the creation of the electronic components is less than 10 μm (ten-millionths of a meter). After the wafers have been sliced from the cylinder, they are polished to a smoothness rivaling the finest mirrors.

The most commonly used fabrication process is *optical lithography* in which ultraviolet light (UV) is passed through a *mask*. The square or rectangular mask carries patterns formed by areas that are either transparent or opaque to ultraviolet frequencies and the resulting image is projected onto the surface of the wafer. The simple patterns shown in the following diagrams were selected for reasons of clarity; in practice a mask can contain hundreds of thousands of fine lines and geometric shapes (Figure 14.2).

Each wafer can contain hundreds or thousands of identical integrated circuits. The pattern projected onto the wafer's surface corresponds to a single integrated circuit, which is typically in the region of 1mm × 1mm to 10mm × 10mm, but may be even larger. After one area has been exposed, the wafer is moved and the process is repeated until the pattern has been replicated across the whole of the wafer's surface. This technique for duplicating the pattern is called a *step-and-repeat* process.

To illustrate the manufacturing process in more detail, we will consider the construction of a single PMOS transistor occupying an area far smaller than a speck of dust. For reasons of electronic stability, the majority of processes begin by lightly doping the entire wafer to form either N-type or, more commonly, P-type silicon. However, for the purposes of this discussion, we will assume a process based on a pure silicon wafer (Figure 14.3).

Ultraviolet radiation source

Mask

Wafer

Figure 14.2: Opto-lithographic step-and-repeat process

Assume that the small area of silicon shown here is sufficient to accommodate a single transistor in the middle of one of the integrated circuits residing somewhere on the wafer. During the fabrication process the wafer is often referred to as the *substrate*, meaning base layer. A common first stage is to either grow or deposit a thin layer of silicon dioxide (glass) across the entire surface of the wafer by exposing it to oxygen in a high-temperature oven (Figure 14.4).

After the wafer has cooled, it is coated with a thin layer of *organic resist* which is first dried and then baked to form an impervious layer (Figure 14.5).

A mask is created and ultraviolet light is applied. The ionizing ultraviolet radiation passes through the transparent areas of the mask into the resist, silicon dioxide, and silicon. The ultraviolet breaks down the molecular structure of the resist, but does not have any effect on the silicon dioxide or the pure silicon (Figure 14.6).

Figure 14.3: Small area of silicon wafer

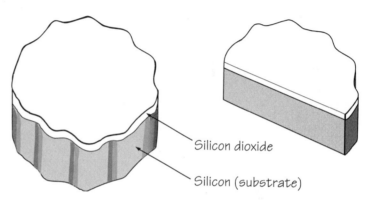

Silicon dioxide

Silicon (substrate)

Figure 14.4: Layer of silicon dioxide grown or deposited

Silicon (substrate)

Silicon dioxide

Organic resist

Figure 14.5: Layer of organic resist applied

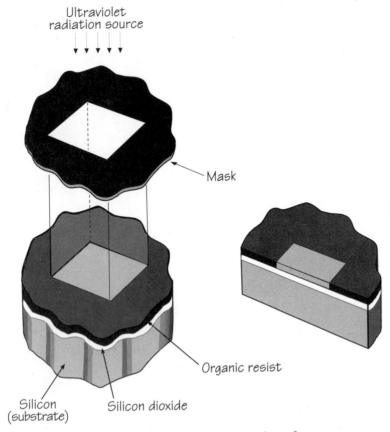

Ultraviolet
radiation source

Mask

Organic resist

Silicon
(substrate)

Silicon dioxide

**Figure 14.6: Exposed resist
degraded by ultraviolet radiation**

Organic resist

Silicon
(substrate)

Silicon dioxide

**Figure 14.7: Degraded resist
dissolved with organic solvent**

As was previously noted, the small area of the mask shown here is associated with a single transistor. The full mask for an integrated circuit would consist of many similar patterns. After the area under the mask has been exposed, the wafer is moved and the process is repeated until the pattern has been replicated across the wafer's entire surface, once for each integrated circuit. The wafer is then bathed in an organic solvent to dissolve the degraded resist. Thus, the pattern on the mask has been transferred to a series of corresponding patterns in the resist (Figure 14.7).

A process in which ultraviolet light passing through the transparent areas of the mask causes the resist to be degraded is known as a *positive-resist* process; *negative-resist* processes are also available. In a negative-resist process, the ultraviolet radiation passing through the transparent areas of the mask is used to cure the resist. The remaining uncured areas are then removed using an appropriate solvent.

After the unwanted resist has been removed, the wafer undergoes a process known as *etching*. An appropriate solvent is used to dissolve any exposed silicon dioxide without having any effect on the organic resist or the pure silicon (Figure 14.8).

The remaining resist is removed and the wafer is placed in a high temperature oven where it is exposed to a gas containing the selected dopant. The atoms in the gas diffuse into the substrate resulting in a region of doped silicon[1] (Figure 14.9).

Organic resist

Silicon (substrate)

Silicon dioxide

Figure 14.8: Etch exposed silicon dioxide

Additional masks and variations on the process are used to create two N-type diffusion regions, a gate electrode, and a layer of insulating silicon dioxide between the substrate and the gate electrode (Figure 14.10).

In the original MOS technologies the gate electrode was metallic—hence the name, metal-oxide semiconductor. In modern processes the gate electrode is formed from *poly-crystalline silicon*, often abbreviated to *polysilicon* or even *poly*, which is also a good conductor.

Gas containing P-type dopant

P-type silicon

Silicon dioxide

Silicon (substrate)

Figure 14.9: Doping exposed silicon

---

[1] In some processes, diffusion is augmented with *ion implantation* techniques in which beams of ions are directed at the wafer to alter the type and conductivity of the silicon in certain regions.

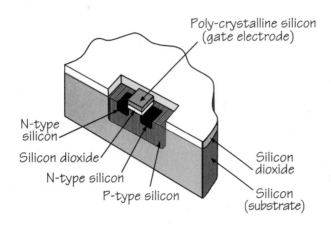

Poly-crystalline silicon
(gate electrode)

N-type
silicon

Silicon dioxide

N-type silicon

P-type silicon

Silicon
dioxide

Silicon
(substrate)

**Figure 14.10: Add gate electrode
and N-type diffusions**

The gap between the source and drain diffusion regions is called the *channel*. To provide a sense of scale, the length of the channel may be in the order of 1 μm (one-millionth of a meter) and the thickness of the silicon dioxide layer between the gate electrode and the substrate may be in the order of 0.05 μm.

Another layer of insulating silicon dioxide is now grown or deposited across the surface of the wafer. Using similar lithographic techniques to those described above, holes are etched through the silicon dioxide in areas in which it is desired to make connections and a metallization layer of aluminum interconnections called *tracks* are deposited (Figure 14.11).

The end result is a PMOS transistor; a logic 0 on the track connected to the gate will turn the transistor ON, thereby

Metal track
(drain)

Metal track
(gate)

Second-layer silicon dioxide

First-layer silicon dioxide

Silicon (substrate)

Metal track
(source)

Figure 14.11: Add metallization

enabling current to flow between its source and drain terminals. An equivalent NMOS transistor could have been created by exchanging the P-type and N-type diffusion regions. By varying the structures created by the masks, components such as resistors, capacitors, and diodes can be fabricated at the same time as the transistors. The tracks are used to connect groups of transistors to form primitive logic gates and to connect groups of these gates to form more complex functions.

An integrated circuit contains three distinct levels of conducting material: the diffusion layer at the bottom, the polysilicon layers in the middle, and the metallization layers at the top. In addition to forming components, the diffusion layer may also be used to create embedded wires. Similarly, in addition to forming gate electrodes, the polysilicon may also be used to interconnect components. There may be several layers of polysilicon and several layers of metallization, each separated by an insulating layer of silicon dioxide. The layers of silicon dioxide are selectively etched with holes known as *vias*, which allow connections to be made between the various tracking layers.

Relatively large areas of aluminum called *pads* are constructed at the edges of each integrated circuit for testing and connection purposes. Some of the pads are used to supply power to the device, while the rest are used to provide input and output signals:

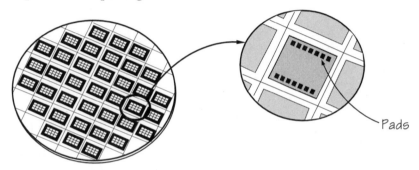

Figure 14.12: Power and signal pads

The pads can be connected to the internal components using the diffusion, polysilicon, or metallization layers. In a step known as *overglassing,* the entire surface of the wafer is coated with a final *barrier layer* (or *passivation layer)* of silicon dioxide or silicon nitride, which provides physical protection for the underlying circuits from moisture and other contaminants. One more lithographic step is required to pattern holes in the barrier layer to allow connections to be made to the pads. In some cases, additional metallization may be deposited on the pads to raise them fractionally above the level of the barrier layer. Augmenting the pads in this way is known as *silicon bumping.*

The entire fabrication process may require numerous lithographic steps, each involving an individual mask and layer of resist to selectively expose different parts of the wafer.

The Packaging Process

The individual integrated circuits are tested while they are still part of the wafer in a process known as *wafer probing*. An automated tester places probes on the device's pads, applies power to the power pads, injects a series of signals into the input pads, and monitors the corresponding signals returned from the output pads. Each integrated circuit is tested in turn, and any device that fails the tests is automatically tagged with a splash of dye for subsequent rejection. The *yield* is the number of devices that pass the tests as a percentage of the total number fabricated on that wafer.

The completed circuits, known as *die*,[2] are separated by marking the wafer with a diamond scribe and fracturing it along the scribed lines (Figure 14.13).

Following separation, the majority of die are packaged individually. Since there are almost as many packaging technologies as there are device manufacturers, we will initially restrain ourselves to a relatively traditional process. First, the die is attached to a metallic *lead frame* using an adhesive (Figure 14.14).

**Figure 14.13:
Die separation**

Lead frame
with die attached

Bare lead frame

Figure 14.14: Die attached to lead frame

---

[2] The plural of die is also die (in much the same way that "a shoal of herring" is the plural of "herring").

One of the criteria used when selecting the adhesive is its ability to conduct heat away from the die when the device is operating. An automatic *wire bonding* tool connects the pads on the die to the leads on the lead frame with wire bonds finer than a human hair. The whole assembly is then encapsulated in a block of plastic or epoxy:

Figure 14.15: Wire bonding and encapsulation

A dimple or notch is formed at one end of the package so that the users will know which end is which. The unused parts of the lead frame are cut away and the device's leads, or pins, are shaped as required; these operations are usually performed at the same time (Figure 14.16).

An individually packaged integrated circuit consists of the die and its connections to the external leads, all encapsulated in the protective package. The package protects the silicon from moisture and other impurities and helps to conduct heat away from the die when the device is operating.

There is tremendous variety in the size and shape of packages. A rectangular device with pins on two sides, as illustrated here, is called a dual-in-line (DIL) package. A standard 14-pin

Figure 14.16: Discard unused lead frame

packaged device is approximately 18 mm long by 6.5 mm wide by 2.5 mm deep, and has 2.5 mm spaces between pins. An equivalent small outline package (SOP) could be as small as 4 mm long by 2 mm wide by 0.75 mm deep, and have 0.5 mm spaces between pins. Other packages can be square and have pins on all four sides, and some have an array of pins protruding from the base. The shapes into which the pins are bent depend on the way the device is intended to be mounted on a circuit board. Additional packaging techniques and alternative mounting strategies are presented in the discussions on circuit boards (Chapter 18), hybrids (Chapter 19), and multichip modules (Chapter 20).

In the early 1990s, some specialist applications began to employ a technique known as *die stacking,* in which several bare die are stacked on top of each other to form a sandwich. The die are connected together and then packaged as a single entity.

Integrated Circuits versus Discrete Components

The tracks linking components inside an integrated circuit have widths measured in millionths of a meter and lengths measured in millimeters. By comparison, the tracks linking components on a circuit board are orders of magnitude wider and have lengths measured in tens of centimeters. Thus, the transistors used to drive tracks inside an integrated circuit can be much smaller than those used to drive their circuit board equivalents, and smaller transistors use less power. Additionally, signals take a finite time to propagate down a track—the shorter the track, the faster the signal.

A single integrated circuit can contain millions of transistors, diodes, resistors, and capacitors. A similar design based on discrete components would be tremendously more expensive in terms of price, size, operating speed, power requirements, and the time and effort required to design and manufacture the system. Additionally, every solder joint on a circuit board is a potential source of failure, which affects the reliability of the design. Integrated circuits reduce the number of solder joints and hence improve the reliability of the system.

Technology Considerations

Transistors are available in a variety of flavors called families or technologies. One of the first to be invented was the bipolar junction transistor (BJT), which has been the mainstay of the industry for many years. If bipolar transistors are connected together in a certain way, the resulting logic gates are classed as transistor-transistor logic (TTL). An alternative method of connecting them results in logic gates classed as emitter-coupled logic (ECL). Metal-oxide semiconductor field-effect transistors (MOS FETs) were invented some time after bipolar junction transistors. Complementary metal-oxide semiconductor (CMOS) logic gates are based on NMOS and PMOS transistors connected together in a complementary manner.

Logic gates constructed in TTL are fast and have strong drive capability, but consume a relatively large amount of power. Logic gates implemented in CMOS are slower than their TTL equivalents and have weaker drive capability, but their static power consumption is extremely low. Technology improvements continue to yield lower-power TTL devices and higher-speed CMOS devices. Logic gates built in ECL are substantially faster than their TTL counterparts but consume correspondingly more power. Finally, gates fabricated using the gallium arsenide (GaAs) semiconductor as a substrate are approximately eight times faster and use one-tenth of the power of their silicon equivalents. However, pure GaAs is difficult to manufacture, hard to work with, and not very robust.

The most complex integrated circuit constructed up to 1994 contained approximately fourteen million transistors. If an integrated circuit of this complexity were constructed entirely from high-power transistors, it could literally consume enough power to incinerate itself (or at least melt itself down into an undistinguished puddle of gunk). As a compromise, some integrated circuits use a combination of technologies. For example, the bulk of the logic gates in a device may be implemented in low-power CMOS, but the gates driving the output pins may be constructed from high-drive TTL. A more extreme example is that of BiCMOS, in which the function of every primitive logic gate is implemented in low-power CMOS, but the output stage of each gate uses high-drive bipolar transistors.

Towards the end of the 1980s and the beginning of the 1990s, the majority of circuits using TTL, CMOS, and BiCMOS devices were based on a 5.0-volt supply. However, increasing usage of portable personal electronics such as notebook computers and cellular telephones began to drive the requirement for devices that consume and dissipate less power. One way to reduce power consumption is to lower the supply voltage and it is anticipated that, by 1996, the most common supplies will be 3.3 volts for portable computers and 3.0 volts for comunication systems. Some specialist applications are even plunging below 2.5 volts. Unfortunately, lowering the supply voltage can drastically affect the speed of traditional technologies and greatly lessens any speed advantages of BiCMOS over CMOS. A relatively new low-voltage contender is BiNMOS, in which complex combinations of bipolar and NMOS transistors are used to form sophisticated output stages providing both high speed and low static power dissipation.

One common metric used to categorize an integrated circuit is the number of logic gates it contains. However, difficulties may arise when comparing devices, as each type of logic function requires a different number of transistors. This leads to the concept of an *equivalent gate*, whereby each type of logic function is assigned an equivalent gate value, and the relative complexity of an integrated circuit is judged by summing its equivalent gates. Unfortunately, the definition of an equivalent gate can vary, depending on whom one is talking to. A reasonably common convention is for a 2-input NAND to represent one equivalent gate. A more esoteric convention defines an ECL equivalent gate as being *"one-eleventh the minimum logic required to implement a single-bit full-adder,"* while some vendors define an equivalent gate as being equal to an arbitrary number of transistors based on their own particular technology. The best policy is to establish a common frame of reference before releasing a firm grip on your hard-earned lucre.

The acronyms SSI, MSI, LSI, VLSI, and ULSI represent Small-, Medium-, Large-, Very-Large-, and Ultra-Large-Scale Integration, respectively. By one convention, the number of gates represented by these terms are: SSI (1–12), MSI (13–99), LSI (100–999) , VLSI (1,000–999,999), and ULSI (1,000,000 or more).

Integrated circuits are also categorized by their *geometries*, meaning the size of the structures created on the substrate. For example, a 1 μm CMOS device has structures that measure one-millionth of a meter. The structures typically embraced by this description are the width of the tracks and the length of the channel between the source and drain diffusion regions; the dimensions of other features are derived as ratios of these structures.

Geometries are continuously shrinking as fabrication processes improve. In 1990, devices with 1 μm geometries were considered to be state of the art. By 1992, many vendors were offering devices with geometries of 0.8 μm, and geometries of 0.6 μm and 0.5 μm were becoming commercially available around the beginning of 1993. It is anticipated that integrated circuits containing twenty million transistors with geometries of 0.35 μm will be technically achievable by 1995, and some industry observers are predicting devices containing one hundred million transistors with geometries of 0.25 μm by the year 2000.

While smaller geometries result in lower power consumption and higher operating speeds, these benefits do not come without a price. Sub-micron logic gates exhibit extremely complex timing effects which make corresponding demands on designers and design systems. Additionally, all materials are naturally radioactive to some extent, and the materials used to package integrated circuits can spontaneously release alpha particles. Devices with smaller geometries are more susceptible to the effects of noise, and the alpha decay in packages can cause corruption of the data being processed by submicron logic gates. There are also some indications that submicron technologies may suffer from a phenomenon known as *subatomic erosion* or, more correctly, *electromigration*, in which the structures in the silicon are eroded by the flow of electrons in much the same way as land is eroded by a river.

Although new techniques are constantly evolving, technologists can foresee the limits of miniaturization that can be practically achieved using optical lithography. These limits are ultimately dictated by the wavelength of ultraviolet radiation. One potential alternative is *X-ray lithography*, but this requires an intense X-ray source and is considerably more expensive than optical lithography. Another possibility is *electron beam lithography*, in which fine electron beams are used to draw extremely high resolution patterns directly into the resist without a mask. Electron beam lithography is sometimes used for custom and prototype devices, but it is much slower and more expensive than optical lithography. Thus, for the present, it would appear that optical lithography will continue to be the mainstay of mass-produced integrated circuits.

Liquid crystal displays are often used for personal electronic appliances such as notebook computers. These displays are arranged as a matrix of points called pixels, each of which can be light or dark, and each of which is controlled by a transistor. Up until now, these displays have typically been fabricated as a sandwich of liquid crystal between two slices of glass. However, the following interesting little snippet was reported in the November 1994 edition of the IEEE Spectrum magazine:

"*Researchers at the French National Center for Scientific Research, in Thiais, say that they have developed an all-plastic, no-metal transistor. Using modern printing techniques, the group built up thin layers of various materials to construct paper-thin field-effect transistors, then employed plastic-like organic polymers and graphite inks to carry electricity. This development could lead to flexible computer screens that roll up like window shades.*"

Personally the author has enough problems with his window shades spontaneously rolling up (which can offer one of life's unforgettable moments when parading in full glory in front of the bathroom mirror) without his computer screens doing the same thing. However, this opens the door to yet more science fiction predictions coming true in the foreseeable future—namely, comic books, newspapers, and textbooks with animated pictures and graphics. We truly do live in interesting times.

Memory ICs

Memory devices are a special class of integrated circuits used to store binary data for later use. There are two main categories of semiconductor memories: read-only memory (ROM) and read-write memory (RWM). Other components in the system can read (extract) data from ROM devices, but cannot write (insert) new data into them. By comparison, data can be read out of RWM devices and, if required, new data can be written back into them. The act of reading data out of a RWM does not affect the master copy of the data stored in the device. For a number of reasons, mainly historical, RWMs are more commonly known as random-access memories (RAMs).

ROM and RAM devices find a large number of diverse applications in electronic designs, but their predominant usage is as memory in computer systems (Figure 15.1). The brain of the computer is the central processing unit (CPU) where all the number crunching is performed. The CPU uses a set of signals called the *address bus* to point to the memory location in which it is interested. The address bus is said to be uni-directional because it only conveys information in one direction: from the CPU to the memory. By means of control signals, the CPU either reads data from, or writes data to, the selected memory location. The data is transferred on a set of signals

Figure 15.1: ROM and RAM in a computer system

called the *data bus*, which is said to be bidirectional because it can convey information in two directions: from the CPU to the memory and vice versa.

The bulk storage is often based on magnetic media such as a disk or tape, which can be used to store a large quantity of information relatively cheaply. Because magnetic media maintains its data when power is removed from the system it is said to be *nonvolatile*. One of the major disadvantages of currently available bulk storage units is their relatively slow speed.[1] The CPU can process data at a much higher rate than the bulk storage can supply or store it. Semiconductor memories are significantly more expensive than bulk storage, but they are also a great deal faster.

ROM devices are *mask-programmable*; that is, the data they contain is hard-coded into them during their construction. ROMs are also classed as being nonvolatile, because their data remains when power is removed from the system. By comparison, RAM devices initialize containing random logic 0 or logic 1 values when power is first applied to a system. Thus, any meaningful data stored inside a RAM must be written into it by other components in the system after it has been powered-up. Additionally, RAMs are said to be *volatile*, because any data they contain is lost when power is removed from the system.

When a computer system is first powered up it doesn't know much about anything. The CPU is hard-wired so that the first thing it does is to read an instruction from a specific memory address—for example, address zero. The components forming the system are connected together in sucn a way that this hard-wired address points to a location in the ROM. The ROM contains a sequence of instructions that are used by the CPU to initialize both itself and other parts of the system. This initialization is known as *boot-strapping* which is derived from the phrase "*pulling himself up by his boot-straps*." At an appropriate point in the initialization, instructions in the ROM cause the CPU to copy a set of master programs known collectively as the *operating system* from the bulk storage into the RAM. Finally, the ROM instructions direct the CPU to transfer its attention to

---

[1] Note the use of the qualifier "*relatively*." Modern bulk storage is actually amazingly fast, but not as fast as the rest of the system.

the operating system instructions in the RAM and the computer is ready for the user to enter the game.

Underlying RAM and ROM Architectures

The smallest unit of memory, called a *cell,* can be used to store a single bit of data—that is, a logic 0 or a logic 1. A number of cells physically grouped together are classed as a *word,* and all the cells in a word are typically written to, or read from, at the same time. The core of a memory device is made up of a number of words arranged as an array:

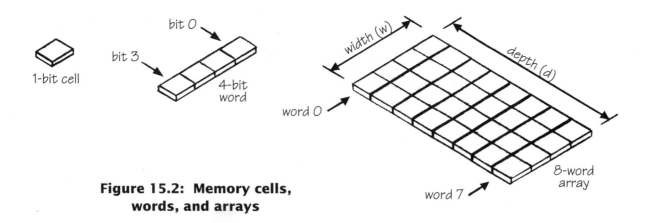

**Figure 15.2: Memory cells,
words, and arrays**

The width (*w*) of a memory is the number of bits used to form a word, where the bits are usually numbered from 0 to (*w* − 1). Similarly, the depth (*d*) of a memory is the number of words used to form the array, where the words are usually numbered from 0 to (*d* − 1).

For the purposes of this discussion it is convenient to visualize a ROM as containing an array of hard-coded cells. In reality, the physical implementation of ROMs is similar to that of PROMs, which are discussed in more detail in Chapter 16.

In *dynamic RAMs* (DRAMs), each cell is formed from a transistor-capacitor pair. The term dynamic is applied because a capacitor loses its charge over time and each cell must be periodically recharged to retain its data. This operation, known as *refreshing,* requires the contents of each cell to be read out and then rewritten. The majority of DRAMs require external circuitry to supervise the refresh process. A special independant controller device is typically employed to manage a group of DRAMs, but some memories with widths of 1 byte or 2 bytes contain their own internal self-refresh circuitry.

In *static RAMs* (SRAMs), each cell is formed from four or six transistors configured as a latch or a flip-flop. The term static is applied because, once a value has been loaded into an SRAM cell, it will remain unchanged until it is explicitly altered or until power is removed from the device.

For external components to reference a particular word in the memory, they must specify that word's *address* by placing appropriate values onto the address_bus. The address is decoded inside the device and the contents of the selected word are made available as outputs from the array. Standard memory devices are constrained to have a depth of 2^n words, where n is the number of bits used to form the address_bus. For example, a 3-bit address_bus can be decoded to select one of eight words ($2^3 = 8$) using a 3:8 decoder:

Figure 15.3: Address bus decoding

This leads to an interesting quirk when referencing the size of a memory device. The prefix k (kilo) represents one thousand, but the closest power of two to one thousand is 2^{10} which equals 1,024. Therefore, a 1 kbit (sometimes shown as "Kbit") memory actually refers to a device containing 1,024 bits. Similarly, the suffix M (mega) is generally taken to represent one million, but the closest power of two to one million is 2^{20} which equals

1,048,576. Therefore, a 1 Mbit memory actually refers to a device containing 1,048,576 bits. If the width of the memory is equal to a byte or a multiple of bytes, then the size of the memory may also be referenced in terms of bytes. For example, a memory containing 1,024 words, each 8 bits wide, may be referred to as being either an 8 kbit or a 1 kbyte device.

Because multiple memory devices are usually connected to a single data bus, the data coming out of the internal array is typically buffered from the external system by means of tri-state gates. Enabling the tri-state gates allows the device to drive data onto the data bus, while disabling them allows other devices to drive the data bus.

In addition to the address bus, a ROM requires a number of control signals, the two most common being ~chip_select and ~read. (The ~read control is sometimes called ~output_enable. Alternatively, some devices have both ~read and ~output_enable controls.) These control signals are commonly active-low; that is, they are considered to be ON when a logic 0 is applied.[2] The ~chip_select signal indicates to the ROM that its attention is required, and it is combined with ~read to form an internal signal ~rd which is used to control the tri-state gates:

Figure 15.4: ROM control decoding and data bus

---

[2] Tilda '~' characters prefixing signal names are used to indicate that these signals are active-low. The use of tilda characters is discussed in detail in Appendix A.

When ~rd is active, the tri-state gates are enabled and the data stored in the word selected by the address_bus is driven out onto the data_bus. When ~rd is inactive, the tri-state gates are disabled and the outputs from the device are placed into high-impedance Z states.

In addition to the control signals used by ROMs, RAMs require a mechanism to control the writing of data into the device. Some RAMs employ separate buses for reading and writing called data_out and data_in, respectively. These components usually have an additional control signal called ~write which is also active-low. Once again, the ~chip_select signal indicates to the RAM that its attention is required, and it is combined with ~read to form an internal ~rd signal which is used to control the tri-state gates. Additionally, ~chip_select is combined with ~write to form an internal ~wr signal (Figure 15.5).

When ~wr is active, the data present on the data_in bus is written into the word selected by the address_bus. The contents of the word pointed to by the address_bus are always available at the output of the array, irrespective of the value on ~wr.

Figure 15.5: RAM with separate data_in and data_out buses

Therefore, devices of this type may be written to and read from simultaneously.

In contrast to those devices with separate data_in and data_out buses, the majority of RAMs use a common bus for both writing and reading. In this case, the ~read and ~write signals are usually combined into a single control input called read~write.[3] When the ~chip_select signal indicates to the RAM that its attention is required, the value on read~write is used to determine the type of operation to be performed. If read~write carries a logic 1, the internal ~rd signal is made active and a read operation is performed; if read~write carries a logic 0, the internal ~wr signal is made active and a write operation is executed:

Figure 15.6: RAM with single bidirectional data bus

When ~rd is active, the tri-state gates are enabled and the data from the word selected by the address_bus is driven onto the data_bus. When ~wr is active (and therefore ~rd is inactive), the tri-state gates are disabled.

---

[3] The name read~write indicates that a logic 1 on this signal is associated with a read operation, while a logic 0 is associated with a write.

This allows external devices to place data onto the data_bus to be written into the word selected by the address_bus.

If the value on the address_bus changes while ~rd is active, the data associated with the newly selected word will appear on the data_bus. However, it is not permissible for the value on the address_bus to change while ~wr is active because the contents of multiple locations may be corrupted.

Increasing Width and Depth

Individual memory devices can be connected together to increase the width and depth of the total memory as seen by the rest of the system. For example, two 1,024-word devices, each 8 bits wide, can be connected so as to appear to be a single 1,024-word device with a width of 16 bits. An address_bus containing 10 bits is required to select between the 1,024 words (Figure 15.7).

The address_bus and control signals are common to both memories, but each device handles a subset of the signals forming the data_bus. Additional devices can be added to further increase the total width of the

Figure 15.7: Connecting devices to increase width

data_bus as required. Alternatively, two 1,024-word devices, each 8 bits wide, can be connected so as to appear to be a single device with a width of 8 bits and a depth of 2,048 words. An address_bus containing 11 bits is required to select between the 2,048 words:

Figure 15.8: Connecting devices to increase depth

The data_bus, the ten least-significant bits of the address_bus, and the read~write control are common to both memories, while the most-significant bit of the address_bus is decoded to generate the ~chip_select signals. A logic 0 on the most-significant bit selects the first device and deselects the second, while a logic 1 deselects the first device and selects the second.

Additional address bits can be used to further increase the total depth as required. If the address_bus in the previous example had contained 12 bits, the two most-significant bits could be passed through a 2:4 decoder to generate four chip selects. These signals could be used to control four 1,024-word devices so as to appear to be a single memory with a width of 8-bits and a depth of 4,096 words.

Alternative Technologies

SRAMs are faster than DRAMs, but each SRAM cell requires more silicon real estate and consumes more power than a corresponding DRAM cell. The first SRAM (256-bit) and DRAM (1,024-bit) devices were both created in 1970. Since that time both types have quadrupled their capacities approximately every three years.[4] By 1993, 4 Mbit SRAMs and 16 Mbit DRAMs were available, pre-release 16 Mbit SRAMs and 64 Mbit DRAMs were undergoing evaluation, and even larger devices such as 256 Mbit DRAMs were being prototyped. Due to their regular structures and the numbers of components produced, memory devices are typically at the leading edge of semiconductor technology by approximately two years. For example, memory devices with 0.35 micron structures are expected to become available by 1994, but other devices with these geometries are not anticipated until 1996.

Because ROMs are constructed containing pre-defined data, they are an expensive design option unless used for large production runs. As an alternative, programmable read-only memories (PROMs) were developed in the mid-1970s. PROMs are manufactured as standard devices that may be electrically programmed by the designer and are nonvolatile. PROMs are one-time programmable and slightly slower than their ROM equivalents, but are significantly cheaper for small- to medium-sized production runs.

To satisfy the designers' desire to combine the advantages of DRAMs (programmable) and ROMs (nonvolatile), erasable programmable read-only memories (EPROMs) were developed in the late 1970s. EPROMs are electrically programmable by the designer and are non-volatile, but can be erased and reprogrammed should the designer so desire. An EPROM device is presented in a ceramic package with a small quartz window mounted in the top. The quartz window is usually covered by a piece of opaque tape. To erase the device, the tape is removed and the EPROM is exposed to ultraviolet radiation for approximately 20 minutes. EPROMs are relatively expensive due to their complicated packaging requirements, but are ideal

---

[4] However, it is predicted that the growth DRAM devices will begin to flatten out, because there is a limit to the amount by which the size of their capacitors can be reduced before they become too small to retain any data.

for prototyping and other applications which require regular changes to the stored data. EPROMs are often used during the development of a design and are subsequently replaced by equivalent PROM or ROM devices in cheaper plastic packages after the design has stabilized. (The actual construction of EPROMs and the other types of memory devices introduced during the remainder of this chapter are discussed in greater detail in Chapter 16.)

A further technology called electrically-erasable read-only memory (EEPROM or E^{2}PROM) was developed towards the end of the 1970s. E^{2}PROMs are electrically programmable by the designer and are non-volatile, but can be electrically erased and reprogrammed should the designer so desire. Additionally, by means of additional circuitry, an E^{2}PROM can be erased and reprogrammed while remaining resident on the circuit board.

An EPROM cell is based on a single transistor and is therefore very efficient in terms of silicon real estate, while an equivalent E^{2}PROM cell is based on two transistors and requires more area on the silicon. However, an EPROM device must be erased and reprogrammed in its entirety, while an E^{2}PROM device can be erased and reprogrammed on a word-by-word basis.

Yet another technology called FLASH is generally regarded as an evolutionary step that combines the best features from EPROM and E^{2}PROM. The name FLASH is derived from its fast reprogramming time compared to EPROM. FLASH has been under development since the end of the 1970s, and was officially described in 1985, but the technology did not initially receive a great deal of interest. However, towards the end of the 1980s, the demand for portable computer and communication systems increased dramatically and FLASH began to attract the attention of designers.

All variants of FLASH are electrically erasable like E^{2}PROMs. Some devices are based on a single transistor cell which provides a greater capacity than an E^{2}PROM, but must be erased and reprogrammed on a device-wide basis similar to an EPROM. Other devices are based on a dual transistor cell and can be erased and reprogrammed on a word-by-word basis. FLASH is considered to be of particular value when the designer requires the ability to reprogram a system in the field or via a communications link while the devices remain resident on the circuit board.

Finally, a class of devices known as nonvolatile RAMs (nvRAMs) may be used in cases where high speed is required and cost is not an overriding issue. These devices are generally formed from an SRAM die mounted in a package with a very small battery, or as a mixture of SRAM and E^2PROM cells fabricated on the same die.

Chapter 16

Programmable ICs

Programmable integrated circuits are standard devices constructed in such a way that a designer can configure them to perform a specific function. These devices, which were introduced in the mid-1970s, are generically known as programmable logic devices (PLDs), and can range in complexity from a few hundred to tens of thousands of equivalent gates. Traditional PLDs are based on logic gates formed from diodes:

Figure 16.1: Diode implementations of AND and OR gates

In the case of the 2-input AND, the pull-up resistor R_y attempts to hold the output y at V_{DD} (logic 1). When both of the inputs are at logic 1, there is no difference in potential across the diodes and y remains at logic 1. However, if input a is connected to V_{SS} (logic 0), diode D_a will conduct and y will be pulled down to logic 0; similarly for input b and diode D_b. Additional inputs can be formed by adding more diodes.

In the case of the 2-input OR, the pull-down resistor R_y attempts to hold the output y at V_{SS} (logic 0). When both of the inputs are at logic 0, there is no difference in potential across the diodes and y remains at logic 0. However, if input a is connected to V_{DD} (logic 1), diode D_a will conduct and y will be pulled up to logic 1; similarly for input b and diode D_b. Once again, additional inputs can be formed by adding more diodes.

In the most common PLD technology, known as *fusible-link*, each diode has an associated link known as a *fuse*. The designer can individually remove unwanted fuses by applying pulses of relatively high voltage and current, in the order of 20 volts and 300 milliamps, respectively, to the device's inputs. These pulses are sufficient to effectively vaporize the fuse:

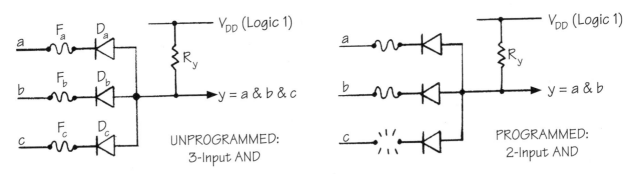

Figure 16.2: Removing unwanted fusible links

This process is typically known as programming the device, but may also be referred to as blowing the fuses or burning the device. Devices based on fusible links are said to be *one-time programmable*, because once a fuse has been blown it cannot be replaced.

As an alternative to fusible links, some PLDs employ antifuse technology in which each diode has an associated *antifuse*. Unlike fusible links which will conduct unless they are forcibly removed, antifuse links will not conduct in the device's unprogrammed state. The user can individually enable desired antifuse links by applying signals of relatively high voltage and current to the device's inputs:

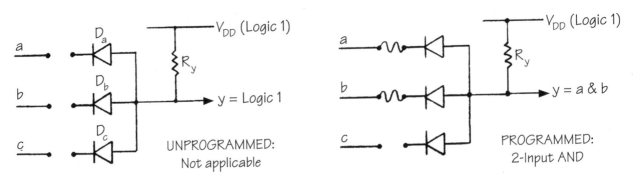

Figure 16.3: Adding desired antifuse links

Antifuse links are formed by creating a connection called a *via* of amorphous (noncrystalline) silicon between two layers of metallization. In its unprogrammed state the amorphous silicon is an insulator with a very high resistance in excess of one billion ohms. The programming signal effectively *grows* a link by changing the insulating amorphous silicon into conducting polysilicon:

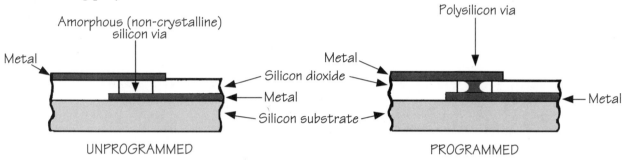

Figure 16.4: Growing an antifuse link

The programmed links in the original antifuse technologies had resistance values in the order of 1,000 ohms, but modern processes have reduced this value to as low as 50 ohms. Devices based on antifuses are also one-time programmable, because once an antifuse link has been grown it cannot be removed.

Due to the fact that PLD structures are very regular, and also because they differ from standard logic gates, a special form of notation has been adopted. This notation is applicable to both fusible-link and antifuse PLDs.

Consider a device delivered with four diode-link pairs forming a 4-input AND function (Figure 16.5).

The AND symbol indicates the function and encompasses the pull-up resistor to V_{DD}. The diode-fuse pairs are represented by the presence or absence of crosses. In a fusible-link device a cross indicates that the corresponding

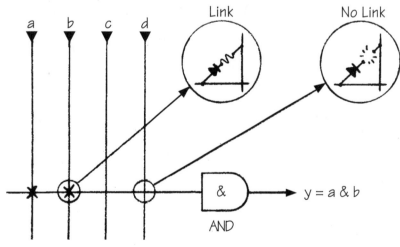

Figure 16.5: Special PLD notation

link has been left intact, while the absence of a cross indicates that the link has been blown away. Thus, the diagram for an unprogrammed fusible-link device will show crosses at each intersection of the matrix. By comparison, in antifuse technology a cross indicates that a link has been grown, while the absence of a cross indicates that no link is present. Thus, the diagram for an unprogrammed antifuse device will not show any crosses.

To increase the versatility of PLDs, their inputs are inverted inside the device and both true and inverse versions of the inputs are presented to an array (Figure 16.6).

The number of AND functions is independent of the number of inputs; additional ANDs can be formed by introducing more rows into the array. Similar techniques can be used to create an array of OR functions, and PLDs typically contain an AND array feeding into an OR array (Figure 16.7).

The number of OR functions is independent of both the number of inputs and the number of AND functions; additional ORs can be formed by introducing more columns into the OR array.

PLDs are not obliged to have AND input arrays feeding OR output arrays; NOR output arrays are also available. However, while it would be possible to create other structures such as OR-AND, NAND-OR, or NAND-NOR, these alternatives tend to be relatively rare or simply not provided. It is not necessary for the vendor to supply all possible variations because AND-OR (and AND-NOR) structures directly map onto the sum-of-products representations most often used to specify equations. Other equation formats can be mapped to these structures using standard Boolean algebraic techniques—for example, DeMorgan transformations.

Although core PLD concepts are relatively simple, the wider arena may prove to be a morass of confusion to the unwary. There are a multiplicity

The majority of this book follows the convention laid down in Appendix A; that is, active-low and complementary signals are indicated by prefixing their names with tilda '~' characters. However, rules are for the guidance of wise men and the blind obedience of fools—the purposes of this chapter are better served by representing inverted PLD input signals by means of lines, or bars, over their names. Actually, this serves to illustrate an important point: although mixing conventions is not usually recommended, engineers are "switch hitters" who use whatever tool is the most appropriate and best suited to get the job done.

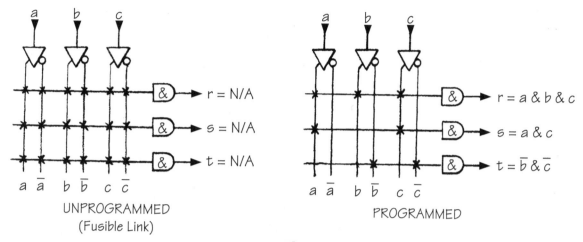

Figure 16.6: True and inverse inputs presented to an array

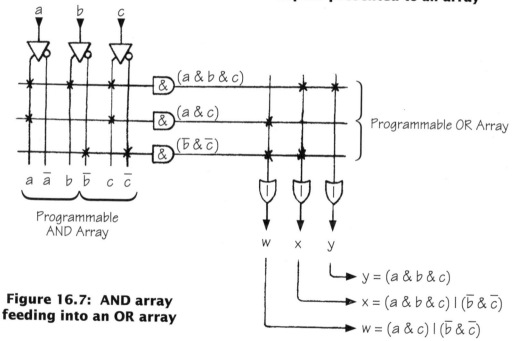

Figure 16.7: AND array feeding into an OR array

of PLD alternatives, most of which seem to have mnemonics formed from different combinations of the same three or four letters. This may be, as some suggest, a strategy to separate the priests from the acolytes, or it may be that the inventors of the devices had no creative energy left to dream up meaningful names for them. Whatever the case, the more common PLD variants are introduced below.

Figure 16.8: PLA
(programmable AND, programmable OR)

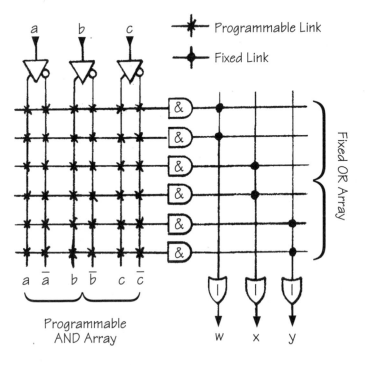

Figure 16.9: PAL
(programmable AND, predefined OR)

Programmable Logic Arrays (PLAs)

The most user-configurable of the traditional PLDs are programmable logic arrays (PLAs), because both the AND and OR arrays are programmable (Figure 16.8).

This diagram indicates a fusible link technology because all the links are present when the device is in its unprogrammed state. Similarly, the following examples are based on fusible link devices unless otherwise stated.

Programmable Array Logic (PAL)

Many applications do not require both the AND and OR arrays to be programmable. In programmable array logic (PAL)[1] devices, only the AND array is programmable while the OR array is predefined (Figure 16.9).

Although PALs are less flexible than PLAs, they operate somewhat faster because hard-wired connections take less time to switch than their programmable equivalents.

---

[1] PAL is a registered trademark of Monolithic Memories Inc.

Programmable Read-Only Memories (PROMs)

In programmable read-only memory (PROM) devices, the AND array is pre-defined and only the OR array is programmable (Figure 16.10). PROMs are useful for equations requiring a large number of product terms, but tend to have fewer inputs because every input combination is decoded and used. The internal structure of a PROM is similar to that of a ROM, the main difference being that a ROM's contents are defined during its construction. PROMs are generally considered to be memory devices where each address applied to the inputs returns a value programmed into the device.

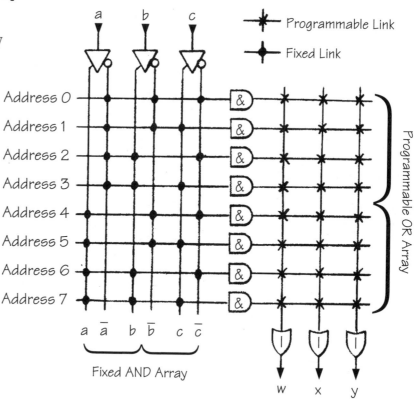

Figure 16.10: PROM (predefined AND, programmable OR)

Additional Programmable Options

Many PLDs have tri-state buffers on their outputs. All of the buffers share a common enable control, which therefore requires only a single input pin on the device (Figure 16.11).

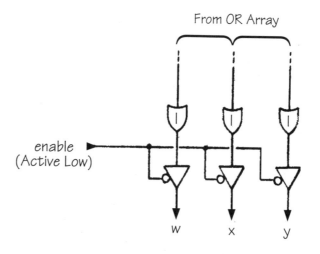

Figure 16.11: PLD with tri-state outputs

Some devices may contain additional programmable links which have not been shown for reasons of clarity. By means of these additional links, each output can be individually programmed to be either permanently enabled or to remain under the control of the *enable* input as required (Figure 16.12).

In this example, the tri-state buffer driving the x output has been disconnected from the *enable* control and connected to a constant logic 0. As a result, the x output is permanently enabled regardless of the value on the *enable* input. To increase the fun, some devices are constructed in such a way that the tri-state outputs are fed back into the AND array (Figure 16.13).

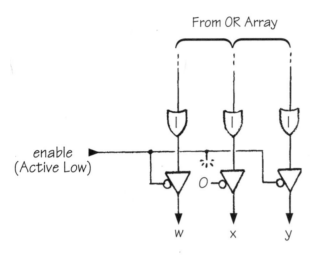

Figure 16.12: PLD with individually tri-statable outputs

Figure 16.13: PLA with tri-state outputs fed back as inputs

Once again there are additional programmable links that have not been shown for reasons of clarity. By means of these links, each output's tri-state buffer can be individually configured so as to always drive a high-impedance Z state. Pins that are configured in this way can be used as additional inputs:

Figure 16.14: PLA with tri-state output configured as an input

In this example, the unprogrammed device began with three inputs and three outputs, but the user actually required a device with four inputs and only two outputs. By means of the appropriate links, the tri-state buffer on the y output was disconnected from the *enable* control and connected to a constant logic 1. Because w and x are still required to function as outputs, their associated links in the AND array must be blown away to ensure that these pins will not have any effect as inputs. The ability to configure pins as outputs or inputs provides a great amount of flexibility; instead of purchasing myriad devices with every conceivable combination of inputs and outputs, the designer can configure a single device type as required.

Certain PLDs are equipped with registers on the outputs, and others with latches. Depending on the particular device, the registers (or latches) may be provided on all the outputs or on a subset of the outputs. Registered devices are particularly useful for implementing finite state machines (FSMs). All of the registers (or latches) share a common control signal which therefore requires only a single input pin on the device (Figure 16.15).

In this example, the outputs are shown as being registered with D-type flip-flops, but alternative register types such as JK flip-flops or T-type flip-flops may be more suitable for certain applications. It can be inconvenient to support a dedicated device for each type of register. As a solution, some devices have configurable register elements whose type can be selected by programming appropriate fuses. Registered (or latched) outputs may also incorporate by-pass multiplexers (Figure 16.16).

By means of appropriate fuses, the control inputs to the multiplexers can be individually configured to select either the nonregistered data or its registered equivalent.

Other common programmable options are the ability to select true or complemented outputs and TTL or CMOS compatible output drivers. An individual PLD typically only provides a subset of the above capabilities, but these may be combined in a variety of ways; for example, registered outputs may be followed by tri-state buffers.

dff= D-type flip-flop

**Figure 16.15: PLD with
registered outputs**

dff = D-type flip-flop
mux = 2:1 multiplexer

**Figure 16.16: Registered outputs
with by-pass multiplexers**

Programming PLDs

Programming a traditional PLD is relatively painless because there are computer programs and associated tools dedicated to the task. The user first creates a computer file, known as a PLD source file, containing a textual description of the required functionality (Figure 16.17).

In addition to Boolean equations, the PLD source file may also support truth tables, state tables, and other constructs, all in textual format. The exclamation marks (called *shrieks*) shown in these equations provide a textual way of indicating inversions. Additional statements allow the user

```
INPUT    =  a , b , c ;
OUTPUT   =  w , x , y ;

EQUATION w  =  (a & b) ;
EQUATION x  =  (a & b) | (a & ! b & ! c) ;
EQUATION y  =  (a & b) | (b & c) | (a & ! c) ;
```

Figure 16.17: PLD source file

to specify which outputs are to be tri-statable, which are to be registered, and any of the other programmable options associated with PLDs.

A special computer program is used to process the PLD source file. This program makes use of a knowledge database that contains details about the internal construction of all the different types of PLDs. After the user has instructed the program as to which type of device they wish to use, it analyzes the equations in the source file and performs algebraic minimization to ensure optimal utilization of the device's resources. The program accesses the knowledge database for details about the designated device and evaluates which fuses need to be blown to implement the desired functionality. The program then generates a textual output file comprising 0 and 1 characters which represent the fuses to be blown; this file is in an industry standard format known as JEDEC (Figure 16.18).

As an alternative to the user specifying a particular device, the program can be instructed to automatically select the best device for the task. The program can base its selection on a variety of criteria such as the speed, cost, and power consumption of the devices. The program may also be used to partition a large design across several devices, in which case it will output a separate JEDEC file for each device.

Knowledge database

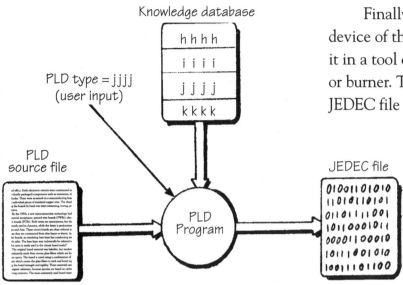

PLD type = jjjj
(user input)

Finally, the designer takes a virgin device of the appropriate type and places it in a tool called a programmer, blower, or burner. The main computer passes the JEDEC file to the burner which uses the contents of the file to determine which fuses to blow. The designer presses the GO button, the burner applies the appropriate signals to the device's inputs, and a new device is born.

Figure 16.18: Programming PLDs

Reprogrammable PLDs

One consideration with fusible link and antifuse technologies is that, once they have been programmed, there is no going back. This may be of particular concern with PROMs as the data they store is prone to change. One alternative is a technology known as erasable programmable read-only memory (EPROM). An EPROM transistor has the same basic structure as a standard MOS transistor, but with the addition of a second polysilicon floating gate isolated by layers of oxide:

Figure 16:19: NMOS transistor versus EPROM transistor

In its unprogrammed state, the floating gate is uncharged and doesn't affect the normal operation of the control gate. To program the transistor, a relatively high voltage in the order of 12V is applied between the control gate and drain terminals. This causes the transistor to be turned hard on, and energetic electrons force their way through the oxide into the floating gate in a process known as hot (high energy) electron injection. When the programming signal is removed, a negative charge remains on the floating gate. This charge is very stable and will not dissipate for more than a decade under normal operating conditions. The stored charge on the floating gate inhibits the normal operation of the control gate, and thus distinguishes those cells that have been programmed from those which have not.

EPROM cells are efficient in terms of silicon real estate, being half the size of DRAM cells and an order of magnitude smaller than fusible links. However, their main claim to fame is that they can be erased and reprogrammed. An EPROM cell is erased by discharging the electrons on the floating gate. The energy required to discharge the electrons is provided by a source of ultraviolet radiation. An EPROM device is delivered in a ceramic package with a small quartz window in the top; this window is usually covered with a piece of opaque sticky tape. To erase the device, it is first removed from the circuit board, the quartz window is uncovered, and the package is placed in an enclosed container with an intense ultraviolet source.

The main problems with EPROM devices are their expensive ceramic packages with quartz windows and the time it takes to erase them, which is in the order of 20 minutes. A foreseeable problem with next generation devices is paradoxically related to improvements in process technology that allows transistors to be made increasingly smaller. The problem is one of scaling, because not all of the structures on a silicon chip are shrinking at the same rate. Most notably, the transistors are shrinking faster than the metal interconnections and, as the features become smaller, a larger percentage of the surface of the die is covered by metal. This makes it difficult for the EPROM cells to absorb the ultraviolet and increases the required exposure time. Some observers feel that it will be impractical to erase EPROMs containing more than 16 megabits.

A somewhat related technology is that of electrically-erasable programmable read-only memory (EEPROM or E^2PROM). An E^2PROM cell is approximately 2.5 times larger than an EPROM cell because it contains two transistors. One of the transistors is similar to that of an EPROM transistor in that it contains a floating gate, but the insulating oxide layers surrounding the floating gate are very much thinner. The second transistor can be used to erase the cell electrically, and E^2PROM devices can typically be erased and reprogrammed on a word-by-word basis.

Finally, a development known as FLASH can trace its ancestry from both EPROM and E^2PROM technologies. The name FLASH was originally coined to reflect the technology's rapid erasure times compared to EPROM. Components based on FLASH can employ a variety of architectures. Some have a single floating gate transistor cell with the same area as an EPROM cell, but with the thinner oxide layers characteristic of an E^2PROM component. These devices can be electrically erased, but only by erasing the whole device or a large portion of it. Other architectures have a two-transistor cell which is almost indistinguishable from that of an E^2PROM cell, allowing them to be erased and reprogrammed on a word-by-word basis.

One advantage of E^2PROM and FLASH devices is that they can be reprogrammed while still on the circuit board. A large proportion of FLASH and E^2PROM devices use hot (high energy) electron injection similar to EPROM and require 12V signals for programming. Some designers see this as an advantage because these devices cannot be accidentally erased by means of the 5V signals used during the circuit's normal operation. However, an increasing number of devices use cold (low energy) Fowler-Nordheim electron tunneling and can be programmed using 5V signals. Device manufacturers are currently working on 3.3V and 3.0V devices to match the anticipated trend in portable computer and telecommunication equipment.

In addition to their use as memories, the EPROM, E^2PROM, and FLASH technologies are also employed in other PLD types which are generically known as EPLDs, E²PLDs, and FLASH PLDs. For example, each fusible link in a standard PLA is replaced by an EPROM cell in an EPLA, an E²PROM cell in an E²PLA, or a FLASH cell in a FLASH PLA. Regardless of the technology used, from fusible link to FLASH, all the components are created on the surface of a single piece of silicon substrate. However, it may be useful to visualize the device as comprising two distinct strata: the AND and OR arrays, and the links:

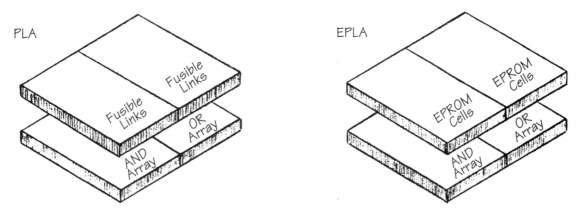

**Figure 16.20: Fusible-link PLA
versus EPLA**

In yet another alternative, some PLDs are constructed with each link being replaced with an SRAM cell. Unlike the other reprogrammable options, PLDs based on SRAM cells are volatile and lose their data when power is removed from the system. However, these devices can be dynamically reprogrammed while the circuit is in use. Some SRAM-based devices employ a double-buffering scheme using two SRAM cells per link. In this case, the device can be reprogrammed with a new pattern while it is still operating with an old pattern. At the appropriate time, the device can be instructed to switch from one bank of SRAM cells to the other.

Reprogrammable devices also convey advantages over fusible link and antifuse devices in that they can be more rigorously tested at the factory. Reprogrammable devices typically undergo one or more program and erase cycles before being shipped to the end user.

Complex PLDs (CPLDs)

The term complex PLD (CPLD) is generally taken to refer to a class of devices that contain a number of PLA or PAL functions sharing a common programmable interconnection matrix (Figure 16.21).

In addition to programming the individual PLD blocks, the connections between the blocks can be configured by means of the programmable interconnection matrix. Additional flexibility may be provided with a CPLD whose links are based on SRAM cells. In this case, it may be possible to configure each PLD block to either act in its traditional role or to function as a block of SRAM.

Programmable interconnect matrix

Figure 16.21: Complex PLD (CPLD)

Field-Programmable Gate Arrays (FPGAs)

Yet another class of devices are known as field-programmable gate arrays (FPGAs). These devices are difficult to categorize because each FPGA vendor fields a proprietary architecture; however, an architecture that illustrates the sophistication of FPGAs could be represented as follows:

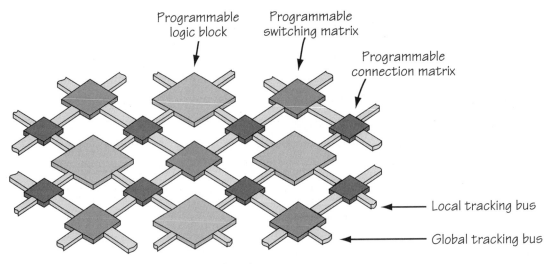

Programmable logic block

Programmable switching matrix

Programmable connection matrix

Local tracking bus

Global tracking bus

Figure 16.22: Field-programmable gate array (FPGA)

The device consists of a number of programmable logic blocks, each connected to a number of programmable connection matrices, which are in turn connected to a number of programmable switching matrices. Each programmable logic block may contain a selection of primitive logic gates and register elements. By programming the appropriate links, each logic block can be individually configured to provide a variety of combinational and/or sequential logic functions. The programmable connection matrices are used to establish links to the inputs and outputs of the relevant logic blocks, while the programmable switch matrices are used to route signals between the various connection matrices. In short, by programming the appropriate links in the connection and switch matrices, the inputs and outputs of any logic block can be connected to the inputs and outputs of any other logic block. Due to their widely varying architectures, FPGA devices typically require vendor specific programming tools.

PLD Applications

A single PLD can be used to replace a large number of simpler devices, thereby reducing board space requirements, optimizing system performance by increasing speed and decreasing power consumption, and reducing the cost of the final system. Additionally, replacing multiple devices with one or more programmable devices increases the reliability of the design. Every device in a design has to be connected to the circuit board by means of soldered joints, and every joint is a potential failure mechanism. Using PLDs reduces the number of devices which reduces the number of joints, thereby reducing the chances of a bad or intermittent connection.

Electronic designs typically undergo a number of changes during the course of their development. A design modification may require the addition or removal of a number of logic functions and corresponding changes to the tracks connecting them. However, when a PLD is used to replace a number of simpler devices, the effects of any design modifications are minimized.

Reprogrammable EPLDs and E^2PLDs are often employed during the initial stages of a design. Once the design has stabilized, these devices may be replaced by cheaper, one-time programmable PLD equivalents. In certain cases involving large production runs, both reprogrammable and one-time programmable devices may eventually be replaced by mask-programmed equivalents, in which the patterns are hard-coded during the construction of the component.

Traditionally, the majority of designers have not availed themselves of the capabilities for in-circuit reprogrammability offered by devices based on the E^2PROM technology. However, the current interest in design-for-test (DFT) and the increasing requirement for portable computer and telecommunication equipment are awakening an interest in in-circuit reprogrammability. This is especially true in the case of PLDs and FPGAs based on FLASH and SRAM technologies which allow devices to be easily and quickly programmed, erased, and reprogrammed while remaining embedded in the middle of a system.

To facilitate the testing of a system, these highly reprogrammable devices can first be configured so as to make internal signals on the circuit board available to the test system. After the circuit has been tested, the devices can be erased and reprogrammed with the patterns required to make the board perform its normal function. In some cases, it may make sense to create a single design that can be used to replace a number of different variants. The designers may create one circuit board type that can be reprogrammed with different features for diverse applications.

Portable electronic equipment such as computers can be reprogrammed in the field or remotely by means of a telephone link. Of most interest is the concept of adaptive hardware which can be reprogrammed on-the-fly. For example, state machines may be reprogrammed to follow different sequences, pattern detectors may be reprogrammed to search for different patterns, and computational devices may be reprogrammed with different algorithms. For all of the reasons discussed above, programmable logic devices form one of the fastest growing segments in the market.

Chapter 17

Application-Specific Integrated Circuits (ASICs)

As opposed to a standard *"off-the-shelf"* integrated circuit whose function is specified by the manufacturer, an application-specific integrated circuit (ASIC) is a device whose function is determined by the designer for a particular application. The programmable logic devices introduced in the previous chapter fall into both camps; in one respect they are standard devices in that they are mass produced, but they may also be classed as subspecies of ASIC because of the designer's ability to customize their functionality. A formal classification of ASICs tends to become a bit fluffy around the edges. However, in addition to programmable logic devices, there are generally taken to be three other major categories: Gate Array (including Sea-of-Gates), Standard Cell (including Compiled Cells), and Full Custom (which also includes Compiled Cells).

Gate Array Devices

A silicon chip may be considered to consist of two major facets: components such as transistors and resistors, and the tracks connecting them together. The most complex, time consuming, and expensive part of a device is the design and fabrication of the components; the addition of the interconnections is relatively less demanding. In the case of gate arrays, the manufacturer prefabricates wafers of silicon chips containing unconnected transistors and resistors.[1] Gate arrays are based on the concept of a *basic cell* which consists of a selection of components; each manufacturer determines

---

[1] In the early days, gate arrays were also known as *uncommitted logic arrays* (ULAs), but this term has largely fallen into disuse.

the numbers and types of components provided in their particular basic cell:

Pure CMOS Basic Cell   BiCMOS Basic Cell

Figure 17.1: Examples of gate array basic cells

In *channeled* gate arrays, the basic cells are typically presented as either single-column or dual-column arrays; the free areas between the arrays are known as the *channels*. Note that, while the diagrams below feature individual silicon chips, at this stage in the process these chips are still embedded in the middle of a wafer:

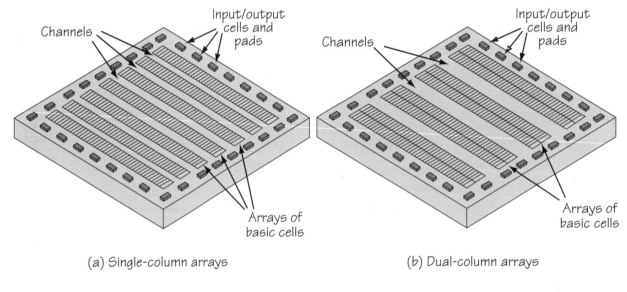

(a) Single-column arrays   (b) Dual-column arrays

Figure 17.2: Channeled gate array architectures

In *channel-less* or *channel-free* devices, the basic cells are presented as a single large array. The surface of the device is covered in a *"sea"* of basic cells and there are no dedicated channels for the interconnections. Thus, these devices are popularly referred to as *sea-of-gates* or *sea-of-cells* (Figure 17.3).

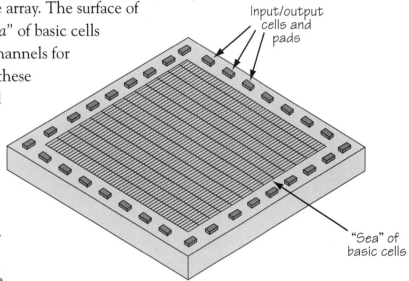

Input/output cells and pads

"Sea" of basic cells

Figure 17.3: Channel-less (sea-of-gates) gate array architecture

Although the transistors from one or more basic cells could be connected together to implement practically any digital function, designers using gate arrays do not work at the transistor level. Each manufacturer selects a set of logic functions such as primitive gates, multiplexers, and registers that they wish to make available to the designer. The manufacturer also determines how each of these functions can be implemented using the transistors from one or more basic cells. A primitive gate may only require the transistors from a single basic cell, while a register may require the use of several basic cells. Each of these "building block" functions is referred to as a *cell* (not to be confused with a basic cell), and the set of functions provided by the manufacturer are known collectively as the *cell library*. The number of functions in a typical cell library can range from 50 to 250 or more. The manufacturer also provides the designer with a cell library data book containing details about each function, including its truth table and timing attributes.

Aided by a suite of design tools, the designer selects the cells and the connections between them that are necessary for the gate array to perform its desired function. The designer passes this information to the manufacturer in a computer readable form, and the manufacturer uses another suite of tools to complete the device. One of these tools assigns the cells selected by the designer to basic cells on the silicon (Figure 17.4).

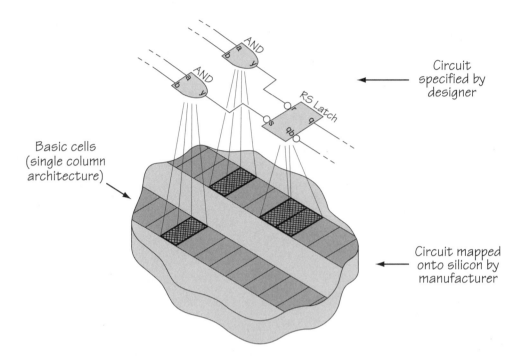

Circuit specified by designer

Basic cells (single column architecture)

Circuit mapped onto silicon by manufacturer

Figure 17.4: Assigning cells to basic cells

Further tools are used to create the masks required to implement the final metallization layers. These layers are used to connect the transistors in the basic cells to form logic gates, and to connect the logic gates to form the complete device. The channels in channeled devices are used for the tracks that connect the logic gates together. By comparison, in channel-less devices the connections between logic gates have to be deposited over the top of other basic cells. With processes based on two layers of metallization, any basic cell overlaid by a track is no longer available to the designer. More recent processes based on three or four metallization layers overcome this problem.

Gate arrays can contain from a few thousand to hundreds of thousands of equivalent gates, but there is a difference between the number of *available gates* and the number of *usable gates*. Due to limited space in the routing channels of channeled devices, the designers may find that they can only actually use between 70% and 90% of the total number of available gates. Sea-of-gates architectures provide significantly more available gates than those of channeled devices because they don't contain any dedicated wiring channels. However, in practice, only about 40% of the gates are usable in devices with two layers of metallization, rising to 60% or 70% in devices with three or four metallization layers.

Functions in gate array cell libraries are generally fairly simple, ranging from primitive logic gates to the level of registers. The device manufacturer may also provide libraries of more complex logical elements called *hard-macros* (or *macro-cells*) and *soft-macros* (or *macro-functions*). In the case of gate arrays, the functions represented by the hard-macros and soft-macros are usually at the MSI and LSI level of complexity, such as comparators, shift-registers, and counters. Hard-macros and soft-macros are both constructed using cells from the cell library. In the case of a hard-macro, the manufacturer predetermines how the cells forming the macro will be assigned to the basic cells and how the connections between the basic cells will be realized. By comparison, in the case of a soft-macro, the assignment of cells to basic cells is performed at the same time, and by the same tool, as for the simple cells specified by the designers.

Gate arrays are classed as *semi-custom* devices. The definition of a semi-custom device is that it has one or more customizable mask layers, but not all the layers are customizable. Additionally, the designer can only utilize the predefined logical functions provided by the manufacturer in the cell and macro libraries.

Standard Cell Devices

Standard cell devices bear many similarities to gate arrays. Once again, each manufacturer decides which logic functions they will make available to the designer. Some manufacturers supply both gate array and standard cell devices, in which case the majority of the logic functions in the cell libraries will be identical and the main differences will be in the timing attributes. Standard cell manufacturers also supply hard-macro and soft-macro libraries which include elements at the LSI and VLSI level of complexity, such as processors, controllers, and communication functions. Additionally, these macro libraries typically include a selection of RAM and ROM functions which cannot be implemented efficiently in gate array devices.

Once again, aided by a suite of design tools, the designers determine which elements they wish to use from the cell and macro libraries, and how they require these elements to be connected together. The designers pass this information to the manufacturer in a computer readable form, and the manufacturer uses another suite of tools to complete the device. However, unlike gate arrays, standard cell devices do not use the concept of a basic cell, and no components are prefabricated on the silicon chip. The manufacturer creates custom masks for every stage of the device's fabrication. This allows each logic function to be created using the minimum number of transistors required to implement that function with no redundant components. Additionally, the cells and macro functions can be located anywhere on the chip, there are no dedicated interconnection areas, and the functions are placed so as to facilitate any connections made between them. Standard cell devices therefore provide a more optimal utilization of the silicon than do gate arrays:

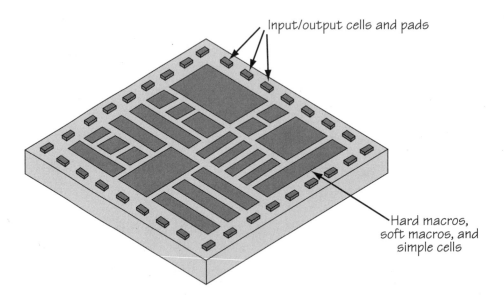

Input/output cells and pads

Hard macros, soft macros, and simple cells

Figure 17.5: Standard cell architecture

Designers of standard cell devices may also make use of *compiled cell* technology. The designer specifies a desired function in such terms as Boolean equations or state tables, and then employs a tool called a *silicon compiler* to generate the masks required to implement the transistors and interconnections. These tools can generate highly efficient structures because they have the ability to vary the size of individual transistors to achieve the optimum tradeoff between speed requirements and power consumption. Silicon compilers are also useful in the creation of selected macro functions such as highly structured RAMs and ROMs. The designer simply specifies the required widths and depths of such functions, leaving the silicon compiler to determine an optimal implementation.

Compiled cell technology was originally intended to provide a mechanism for the creation of entire devices. This promise was never fully realized, and the technology is now considered to be an adjunct to standard cell devices, allowing the designer to create custom cells and macros to augment the libraries supplied by the manufacturer.

From the manufacturer's point of view, standard cell devices may be considered to be *full-custom* on the basis that all the mask layers are customized. However, from the designer's point of view, these devices are considered to be semi-custom because, with the exception of compiled cells, the designer can only utilize the predefined functions provided by the manufacturer in the cell and macro libraries.

Full Custom Devices

In the case of full custom devices, the designer has complete control over every mask layer used to fabricate the silicon chip. The manufacturer does not prefabricate any components on the silicon and does not provide any cell libraries. With the aid of appropriate design tools, designers can *"handcraft"* the dimensions of individual transistors, and then create higher-level functions based on these components. For example, if the designers require a slightly faster logic gate, they can alter the dimensions of the transistors used to build that gate. The design tools used for full custom devices are often created in-house at the designer's facility.

Full custom devices may employ compiled cell technology, and may also include analog circuitry such as comparators, amplifiers, filters, and digital-to-analog and analog-to-digital converters. The design of full custom devices is highly complex and time consuming, but the resulting chips contain the maximum amount of logic with minimal waste of silicon real estate.

Input/Output Cells and Pads

Around the periphery of the silicon chip are power and signal pads used to interface the device to the outside world. The cell library data books for gate array and standard cell devices include a set of functions known as input/output (I/O) cells. The signal pads contain a selection of transistors, resistors, and diodes necessary to implement input, output, or bidirectional buffers, and the designer can decide how each pad will be configured. The masks and metallization used to interconnect the internal logic are also used to configure the components in the input/output cells and to connect the internal logic to these cells (Figure 17.6).

The manufacturer may permit the designer to individually specify whether each input/output cell should present CMOS, TTL or ECL characteristics to the outside world. The input/output cells also contain any circuitry required to provide protection against electrostatic discharge (ESD). After the metallization layers have been added, the chips are separated and packaged using the same techniques as for standard integrated circuits.

Figure 17.6: Input/output cells and pads

Applications and Technology Considerations

Electronic designs typically contain a selection of highly complex functions such as processors, controllers, and memory, interfaced to each other by a plethora of simple functions such as primitive logic gates. The simple interfacing functions are often referred to as the *glue logic*. Both complex and simple functions are available as standard integrated circuits. Devices containing small numbers of simple logic functions (for example, four 2-input AND gates) are sometimes referred to as *jelly bean* parts. When glue logic is implemented using jelly bean parts, it may require a disproportionate amount of the circuit board's real estate. A few complex devices containing the vast majority of the design's logic gates may occupy a small area, while a large number of jelly bean devices containing relatively few logic gates can occupy the bulk of the board's surface.

Early ASICs contained only a few hundred logic gates. These were mainly used to implement glue logic and it was possible to replace fifty or more jelly-bean devices with a single ASIC. This greatly reduced the size of circuit boards, increased speed and reliability, and reduced power consumption. As time progressed, application-specific devices with hundreds of thousands of gates and hundreds of pins became available. Thus, in addition to glue logic, today's ASICs can be used to implement the most complex functions.

Both gate array and standard cell devices essentially consist of building blocks designed and characterized by the manufacturer and connected together by the designer. Gate arrays are mask programmable with a predefined number of transistors, and different designs are essentially just changes in the interconnect. Gate arrays only require the customization of around four mask layers as opposed to around fourteen for standard cell devices. Gate arrays are therefore faster to prototype than their standard cell equivalents, but the latter can contain more logic gates.

Gate arrays are moving towards having similar capabilities to standard cells, and some support complex functions such as memory and processor cores. The designer generally knows the memory and processing requirements in advance of the rest of the logic, and the gate array manufacturer may supply devices with pre-built processor and memory functions surrounded by arrays of basic cells. Some gate array devices support simple analog cells in addition to the digital cells, and standard cell devices can be constructed with complex analog functions if required.

ASICs, CSICs, or ASSPs?

Finally, although the term application-specific integrated circuit (ASIC) is almost a de-facto standard, some observers believe that this name has been incorrect from the beginning. There is some justification for this point of view because "application-specific" does tend to imply that the resulting device is only applicable to one particular application, but some of these devices find use in many applications. For this reason, the more pedantic in our midst would prefer to say customer-specific integrated circuit (CSIC), but this term is rarely used in the industry and shows little indication of finding favor with the masses.

However, there is an increasing trend for device manufacturers to create integrated circuits using ASIC technologies, and for these devices to be sold as standard parts to anybody who wants to buy them. This has led to a relatively new term that appears to be gaining in popularity: application-specific standard part (ASSP).

Chapter 18

Circuit Boards (PWBs and DWBs)

Electronic components are rarely useful in isolation, and it is usually necessary to connect a number of them together in order to achieve a desired effect. Early electronic circuits were constructed using discrete (individually packaged) components such as transistors, resistors, capacitors, and diodes. These were mounted on a nonconducting board and connected using individual pieces of insulated copper wire. The thankless task of wiring the boards by hand was time-consuming, boring, prone to errors, and expensive.

By the 1950s, a new interconnection technology had gained commercial acceptance: printed wire boards (PWBs), also known as printed circuit boards (PCBs). Both terms are synonymous, but the former is more commonly used in America, while the latter is predominantly used in Europe and Asia. These circuit boards are often referred to as *laminates* because they are constructed from thin layers or sheets. In the case of the simpler boards, an insulating base layer has conducting tracks formed on one or both sides. The base layer may technically be referred to as the *substrate*, but this term is rarely used in the circuit board world.[1]

The original board material was bakelite, but modern boards are predominantly made from woven glass fibers which are bonded together with an epoxy. The board is cured using a combination of temperature and pressure which causes the glass fibers to melt and bond together, thereby giving the board strength and rigidity. These materials may also be referred to as *organic substrates*, because epoxies are based on carbon compounds as are living creatures. The most commonly used board material of this type is

---

[1] See also the glossary definition of "substrate."

known as *FR4,* where the first two characters stand for flame retardant, and you can count the number of people who know what the "4" stands for on the fingers of one hand.

To provide a sense of scale, a fairly representative board might be 15 cm × 20 cm in area and in the region of 1.5 mm to 2.0 mm thick, but they can range in size from 2 cm × 2 cm or smaller (and thinner) to 50 cm × 50 cm or larger (and thicker).[2]

Subtractive Processes

In a subtractive process, a thin layer of copper foil in the order of 0.02mm thick is bonded to the surface of the board. The copper's surface is coated with an *organic resist* which is cured in an oven to form an impervious layer:

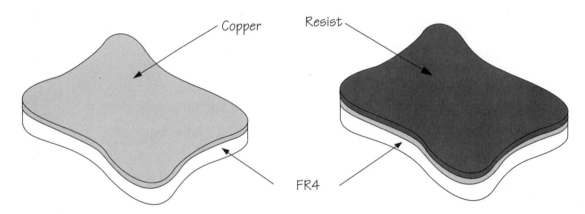

**Figure 18.1: Subtractive processes:
copper clad board**

Next, an optical mask is created with areas that are either transparent or opaque to ultraviolet light. The mask is usually the same size as the board, and can contain hundreds of thousands of fine lines and geometric shapes.

---

[2] These dimensions are not intended to imply that circuit boards must be square or even rectangular. In fact, circuit boards may be constructed with whatever outline is necessary to meet the requirements of a particular enclosure; for example, the shape of a car dashboard.

The mask is placed over the surface of the board which is then exposed to ultraviolet light. This ionizing radiation passes through the transparent areas of the mask to break down the molecular structure of the resist. After the board has been exposed, it is bathed in an organic solvent which dissolves the degraded resist. Thus, the pattern on the mask has been transferred to a corresponding pattern in the resist:

Figure 18.2: Subtractive processes: exposing the resist

A process in which ultraviolet light passing through the transparent areas of the mask causes the resist to be degraded is known as a *positive-resist* process; *negative-resist* processes are also available.[3] The following discussions assume positive-resist processes unless otherwise noted.

After the unwanted resist has been removed, the board is placed in a bath containing a cocktail based on sulfuric acid, which is agitated and aerated to make it more active. The sulfuric acid dissolves any exposed

---

[3] In a negative-resist process the ultraviolet radiation passing through the transparent areas of the mask is used to cure the resist. The remaining uncured areas are then removed using an appropriate solvent. Thus, a mask used in an negative-resist process is the photographic negative of one used in a positive-resist process to achieve the same effect; that is, the transparent areas are now opaque and vice versa.

copper not protected by the resist in a process known as *etching*. The board is then washed to remove the remaining resist. Thus, the pattern in the mask has now been transferred to a corresponding pattern in the copper:

**Figure 18.3: Subtractive processes:
removing the unwanted copper**

This type of process is classed as subtractive because the board is first covered with the conductor and any unwanted material is then removed, or subtracted. As a point of interest, much of this core technology predates the modern electronics industry. The process of copper etching was well known by the printing industry in the 1800s, and opto-lithographic techniques involving organic resists were used to create printing plates as early as the 1920s. These existing processes were readily adopted by the fledgling electronics industry. As process technologies improved it became possible to achieve ever finer features. By the early 1990s, a large proportion of boards were using *lines* and *spaces* of 0.12 mm. The term lines refers to the widths of the tracks and the term spaces refers to the gaps between adjacent tracks. By the mid-1990s it is anticipated that mainstream boards will be using lines and spaces of 0.08 mm.[4]

---

[4] In fact, leading-edge boards were already being manufactured with significantly finer geometries than these as early as 1993.

Additive Processes

An additive process does not involve any copper foil being bonded to the board. Instead, the coating of organic resist is applied directly to the board's surface:

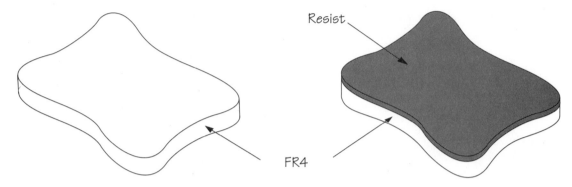

**Figure 18.4: Additive processes:
bare board**

Once again the resist is cured, an optical mask is created, ultraviolet light is passed through the mask to break down the resist, and the board is bathed in an organic solvent to dissolve the degraded resist:[5]

**Figure 18.5: Additive processes:
exposing the resist**

---

[5] Note that a mask used in an additive process is the photographic negative of one used in a subtractive process to achieve the same effect; that is, the transparent areas are now opaque and vice versa.

After the unwanted resist has been removed, the board is placed in a bath containing a cocktail based on copper sulfate where it undergoes a process known as *electroless plating*. Tiny crystals of copper grow on the exposed areas of the board to form copper tracks. The board is then washed in an appropriate solvent to remove the remaining resist:

Undegraded resist

Copper grown on FR4

Remaining resist removed to leave bare FR4

Copper

Figure 18.6: Additive processes: adding the desired copper

A process of this type is classed as additive because the conducting material is only grown on, or added to, specific areas of the board. Additive processes are increasing in popularity because they require less processing and result in less wasted material. Additionally, fine tracks can be grown more accurately in additive processes than they can be etched in their subtractive counterparts. These processes are of particular interest for high-speed designs and microwave applications, in which conductor thicknesses and controlled impedances are critical. Groups of tracks, individual tracks, or portions of tracks can be built up to precise thicknesses by iterating the process multiple times with selective masking.

Single-sided Boards

It probably comes as no great surprise to find that single-sided boards have tracks on only one side. These tracks, which may be created using either subtractive or additive processes, are terminated with areas of copper known as *pads*. The shape of the pads and other features are, to some extent, dictated by the method used to attach components to the board. Initially, these discussions will assume that the components are to be attached using a technique known as *through-hole*, which is described in more detail below. The pads associated with the through-hole technique are typically circular, and holes are drilled through both the pads and the board using a computer-controlled drilling machine:[6]

**Figure 18.7: Single-sided boards:
drilling the holes**

Once the holes have been drilled, an electroless plating process referred to as *tinning* is used to coat the tracks and pads with a layer of tin-lead alloy. This alloy is used to prevent the copper from oxidizing and provides protection against contamination. The deposited alloy has a rough surface composed of vertical crystals called *spicules* which, when viewed under a microscope, resemble a bed-of-nails. To prevent oxygen from reaching the copper through pin holes in the alloy, the board is placed in a *reflow oven* where it is heated by either infrared (IR) radiation or hot air. The reflow oven causes the alloy to melt and form a smooth surface (Figure 18.8).

---

[6] This is usually referred to as an NC drilling machine, where NC stands for "Numerically Controlled."

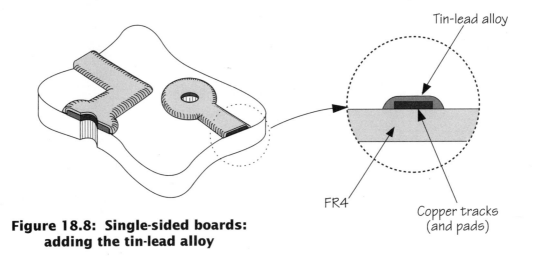

Figure 18.8: Single-sided boards: adding the tin-lead alloy

After the board has cooled, a layer known as the *solder mask* is applied to the surface carrying the tracks. One common technique is for the solder mask to be screen printed onto the board. In this case, the screen used to print the mask has patterns that leave the areas around the pads exposed, and the mask is then cured in an oven. In an alternative technique, the solder mask is applied across the entire surface of the board as a film with an adhesive on one side. In this case, a further opto-lithographic stage is used to cure the film with ultraviolet light. The optical mask in this process contains opaque patterns which prevent the areas around the pads from being cured; these areas are then removed using an organic solvent:

Figure 18.9: Single-sided boards: adding the solder mask

Beware! Although there are relatively few core processes used to manufacture circuit boards, there are almost endless variations and techniques. For example, the tracks and pads may be created before the holes are drilled or vice versa. Similarly, the tin-lead alloy may be applied before the solder mask or vice versa. This latter case, known as *solder mask over bare copper (SMOBC)*, prevents solder from leaking under the mask when the tin-lead alloy melts during the process of attaching components to the board. Thus, the tin-lead alloy is only applied to any areas of copper which are left exposed by the solder mask, such as the pads at the end of the tracks. As there are so many variations and techniques, these discussions can only hope to offer an overview of the main concepts.

Lead Through-Hole (LTH)

Prior to the early 1980s, the majority of integrated circuits were supplied in packages that were attached to a circuit board by inserting their leads through holes drilled in the board. This technique, which is still widely used, is known as lead through-hole (LTH), plated through-hole (PTH), or, more concisely, *through-hole*. In the case of a single-sided board, any components that are attached to the board in this fashion are mounted on the opposite side to the tracks. This means that any masks used to form the tracks are actually created as mirror-images of the required patterns:

Figure 18.10: Lead through-hole (LTH)

The act of attaching components is known as *populating* the board, and the area of the board occupied by a component is known as its *footprint*.[7] Early manufacturing processes required the boards to be populated by hand, but modern processes make use of automatic insertion machines. Similarly, component leads used to be hand-soldered to the pads, but most modern processes employ automatic *wave soldering* machines:

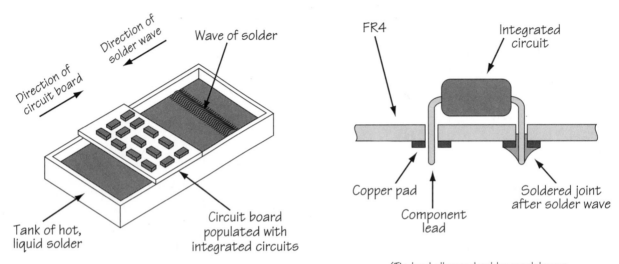

Figure 18.11: Wave soldering

A wave soldering machine is based on a tank containing hot, liquid solder.[8] The machine creates a wave (actually a large ripple) of solder which travels across the surface of the tank. Circuit boards populated using the through-hole technique are passed over the machine on a conveyer belt. The system is carefully controlled and synchronized such that the solder wave brushes across the bottom of the board only once.

The solder mask introduced in the previous section prevents the solder from sticking to anything except the exposed pads and component leads. Because the solder is restricted to the area of the pads, surface tension causes

---

[7] Actually, the term "footprint" is primarily used in the context of surface mount technology, which is introduced below. However, this term may be used to describe the area occupied by a device using any population strategy.

[8] An alloy of tin and lead with a comparatively low melting point.

it to form good joints between the pads and the component leads. Additionally, capillary action causes the solder to be drawn up the hole, thereby forming reliable, low-resistance connections. If the solder mask were omitted, the solder would run down the tracks away from the component leads. In addition to forming bad joints, the amount of heat absorbed by the tracks would cause them to separate from the board (this is not considered to be a good thing to happen).

Surface Mount Technology (SMT)

In the early 1980s, new techniques for packaging integrated circuits and populating boards began to appear. One of the more popular is surface mount technology (SMT), in which component leads are attached directly to pads on the surface of the board. Components with packages and lead shapes suitable for this technology are known as surface mount devices (SMDs). The most commonly used package to achieve a high lead count in a small area is the quad flat pack (QFP), in which leads are present on all four sides of a thin square package.[9]

Boards populated with surface mount devices are fabricated in much the same way as their through-hole equivalents (except that the pads used for attaching the components are typically square or rectangular and do not have any holes drilled through them). However, the processes begin to diverge after the solder mask and tin-lead plating layers have been applied. A layer of solder paste is screen printed onto the pads, and the board is populated by an automatic pick-and-place machine which pushes the component leads into the paste (Figure 18.12).

Thus, in the case of a single-sided board, the components are mounted on the same side as the tracks. When all of the components have been attached, the solder paste is melted to form good conducting bonds between their leads and the board's pads. The solder paste can be melted by placing the board in a *reflow oven* where it is heated by infrared radiation or hot air. Alternatively, the solder may be melted using *vapor-phase soldering*, in which

---

[9] Other packaging styles such as pad grid arrays and ball grid arrays, which are also suitable for surface mount technology, are introduced in more detail in Chapter 20.

Integrated circuit
(quad flat package)

Component lead

Solder paste

Pad

FR4

(Tin-lead alloy and solder mask layers
omitted for reasons of clarity)

Tracks

No holes in pads

**Figure 18.12: Surface mount
technology (SMT)**

the board is lowered into the vapor-cloud above a tank containing boiling hydrocarbons. However, vapor-phase soldering is becoming increasingly less popular due to environmental concerns.

Surface mount technology is well suited to automated processes. Because the components are attached directly to the surface of the board, they can be constructed with leads that are finer and more closely spaced than their through-hole equivalents. The result is smaller and lighter packages which can be mounted closer together and occupy less of the board's surface area which is referred to as *real estate*. This, in turn, results in smaller, lighter, and faster circuit boards. The fact that the components do not require holes for their leads is also advantageous, because drilling holes is a time-consuming and expensive process. Additionally, if any holes are required to make connections through the board (as discussed below), they can be made much smaller because they do not have to accommodate component leads.[10]

---

[10] Typical pad and hole diameters are presented in the *"Holes versus Vias"* section later in this chapter.

Double-sided Boards

There is a simple game played by children all over the world. The game commences by drawing three circles and three squares on a piece of paper, and then trying to connect each circle to each square without any of the connecting lines crossing each other (Figure 18.13).

Children can devote hours to this game, much to the delight of their parents. Unfortunately, there is

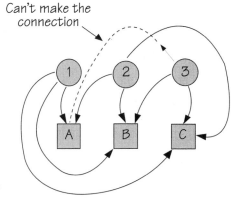

Figure 18.13: Circles and squares game

no solution, and one circle-square pair will always remain unconnected. This simple example illustrates a major problem with single-sided boards, which may have to support thousands of component leads and tracks. If any of the tracks cross, an undesired electrical connection will be made and the circuit will not function as desired. One solution to this dilemma is to use wire links called *jumpers:*

(Tin-lead alloy and solder mask layers omitted for reasons of clarity)

Figure 18.14: Single-sided boards: using jumpers

Unfortunately, the act of inserting a jumper is as expensive as for any other component. At some point it becomes more advantageous to employ a double-sided board which has tracks on both sides.

Initially, the construction of a double-sided board is similar to that for a single-sided board. Assuming a subtractive process, copper foil is bonded to both sides of the board, then organic resist is applied to both surfaces and cured. Separate masks are created for each side of the board and ultraviolet light is applied to both sides. The ultraviolet radiation that is allowed to pass through the masks degrades the resist which is then removed using an organic solvent. Any exposed copper that is not protected by resist is etched, the remaining resist is removed, and holes are drilled. However, a double-sided board now requires an additional step. After the holes have been drilled, a plating process is used to line them with copper (Figure 18.15).

Instead of relying on jumpers, a track can now pass from one side of the board to the other by means of these copper-plated holes known as *vias*.[11,12] The tracks on one side of the board usually favor the y-axis (north-south), while the tracks on the other side favor the x-axis (east-west). The inside of

Figure 18.15: Double-sided boards: creating vias

---

[11] The term via is taken to mean a conducting path linking two or more conducting layers, but does not include a hole accommodating a component lead (see also the following section, "*Holes versus Vias*").

[12] There are a number of alternative techniques which may be used to create circuit board vias but, by default, the term is typically understood to refer to holes plated with copper as described here.

the vias and the tracks on both sides of the board are plated with tin-lead alloy, and solder masks are applied to both surfaces (or vice versa in the case of SMOBC processes).

Some double-sided boards are populated with through-hole or surface mount devices on only one side. Some boards may be populated with through-hole devices on one side and surface mount devices on the other. And some boards may have surface mount devices attached to both sides. This latter case is of particular interest, because surface mount devices do not require holes to be drilled through the pads used to attach their leads (they have separate fan-out vias as discussed below). Thus, in surface mount technology, it is possible to place two devices directly facing each other on opposite sides of the board without making any connections between them.

Having said this, in certain circumstances it may be advantageous to form connections between surface mount devices directly facing each other on opposite sides of the board. The reason for this is that a through-board connection can be substantially shorter than an equivalent connection between adjacent devices on the same side of the board. Thus, this technique may be of use for applications such as high-speed data buses, because shorter connections result in faster signals.

Holes versus Vias

Manufacturers of circuit boards are very particular about the terminology they use, and woe betide anyone caught mistakenly referring to a hole as a via or vice versa. Figure 18.16 should serve to alleviate some of the confusion.

In the case of a single-sided board (as illustrated in Figures 18.7 and 18.14), a hole that is used to accommodate a through-hole component lead is simply referred to as a *hole*. By comparison, in the case of double-sided boards (or multilayer boards as discussed below), a hole that is used to accommodate a through-hole component lead is plated with copper and is referred to as a *plated through-hole*. Additionally, a hole that is only used to link two or more conducting layers, but does not accommodate a component lead, is referred to as a *via* (or, for those purists among us, an *interstitial via*). The qualification attached to this latter case is important, because even if a hole that is used to

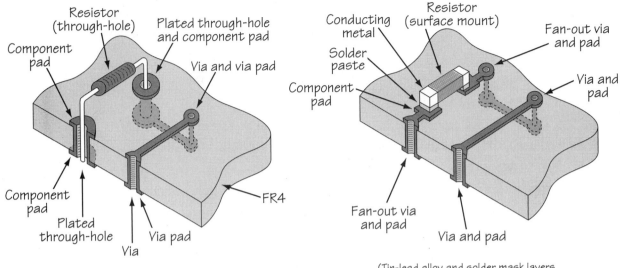

Figure 18.16: Holes versus vias

(Tin-lead alloy and solder mask layers
omitted for reasons of clarity)

accommodate a component lead is also used to link two or more conducting layers, it is still referred to as a plated through-hole and not a via.

Due to the fact that vias do not have to accommodate component leads, they can be created with smaller diameters than plated through-holes, thereby occupying less of the board's real estate. To provide a sense of scale, the diameters of plated through-holes and their associated pads are usually in the order of 0.6 mm and 1.2 mm, respectively, while the diameters of vias and via pads can be less than 0.25 mm and 0.5 mm, respectively.

Finally, in the case of surface mount devices attached to double-sided boards (or multilayer boards as discussed below), each component pad is usually connected by a short length of track to a via, known as a *fan-out via*,[13] which forms a link to other conducting layers. However, if this track exceeds a certain length (it could meander all the way around the board), an otherwise identical via at the end would be simply referred to as a via. Unfortunately, there is no standard length of track that differentiates a fan-out via from a standard via, and any such classification depends solely on the in-house design rules employed by the designer and board manufacturer.

---

[13] Some engineers attempt to differentiate vias that fall inside the device's footprint (under the body of the device) from vias that fall outside the device's footprint by referring to the former as *fan-in vias*, but this is not a widely used term.

Multilayer Boards

It is not unheard of for a circuit board to support thousands of components and tens of thousands of tracks and vias. Double-sided boards can support a higher population density than single-sided boards, but there quickly comes a point when even a double-sided board reaches its limits. A common limiting factor is lack of space for the necessary number of vias. To overcome this limitation, designers may move onwards and upwards to multilayer boards.

A multilayer board is constructed from a number of single-sided or double-sided sub-boards.[14] The individual sub-boards can be very thin, and multilayer boards with four or six conducting layers are usually around the same thickness as a standard double-sided board. Multilayer boards may be constructed using a double-sided sub-board at the center with single-sided sub-boards added to each side.[15] Alternatively, they may be constructed using only double-sided sub-boards separated by non-conducting layers of semi-cured FR4 known as *prepreg*:

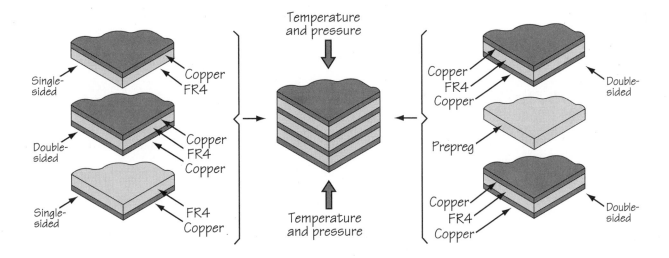

**Figure 18.17: Multilayer boards:
alternative structures**

---

[14] The term "sub-board" is not an industry standard, and it is only used in these discussions to distinguish the individual layers from the completed board.

[15] This technique is usually reserved for boards which only carry four conducting layers.

After all of the layers have been etched to form tracks and pads, the sub-boards and prepreg are bonded together using a combination of temperature and pressure. This process also serves to fully cure the prepreg. The majority of multilayer boards have less than ten conducting layers, but boards with twenty-four conducting layers or more are not outrageously uncommon.

To overcome the problem of limited space, multilayer boards may use *through-hole, blind,* and *buried vias.* A through-hole via passes all the way through the board, a blind via is only visible from one side of the board, and a buried via is used to link internal layers and is not visible from either side of the board (Figure 18.18).

Unfortunately, although they help to overcome the problem of limited space, blind and buried vias significantly increase the complexity of the manufacturing process. When these vias are only used to link both sides of a sub-board, that board must be drilled individually and a plating process used to line its vias with copper. Similarly, when these vias only pass through a number of sub-boards, those boards must be bonded together, drilled, and plated as a group. Finally, after all of the sub-boards have been bonded together, any holes that are required to form plated through-holes and vias are drilled and plated. Blind and buried vias can greatly increase the number of tracks that a board can support but, in

Cross-sectional view

Through-hole via

Blind via

Buried via

Layers of FR4

(Tin-lead alloy and solder mask layers omitted for reasons of clarity)

**Figure 18.18: Multilayer boards:
through-hole, blind, and buried vias**

addition to increasing costs and fabrication times, they can also make it an absolute bugger to test.

The layers carrying tracks are known as the *signal layers*. In a multilayer board, the signal layers are typically organized so that each pair of adjacent layers favor the y-axis and the x-axis, respectively. Additionally, two or more conducting layers are typically set aside to be used as power and ground planes. The power and ground planes usually occupy the central layers, but certain applications have them on the board's outer surfaces. This latter technique introduces a number of problems, but it also increases the board's protection from external sources of *noise* such as electromagnetic radiation.

Unlike the signal layers, the bulk of the copper on the power and ground planes remains untouched. The copper on these layers is only etched away in those areas where it is not required to make a connection. For example, consider a through-hole device with eight leads. Assume that leads 4 and 8 connect to the ground and power planes respectively, while the remaining leads are connected into various signal layers (Figure 18.19).

For the sake of simplicity, the exploded view only shows the central sub-board carrying the power planes; the sub-boards carrying the signal layers would be bonded to either side. Also note that the holes shown in the prepreg in the exploded view would not be drilled and plated until all of the sub-boards had been bonded together.

In the case of component leads 1, 2, 3, 5, 6, and 7, both the power and ground planes have copper removed around the holes. These etched-away areas, known as *anti-pads*, are used to prevent connections to the planes when the holes are plated. Similarly, the power plane has an anti-pad associated with lead 4 (the ground lead), and the ground plane has an anti-pad associated with lead 8 (the power lead).

The power plane has a special pattern etched around the hole associated with lead 8 (the power lead), and a similar pattern is present on the ground plane around the hole associated with lead 4 (the ground lead). These patterns, known as *thermal relief pads*,[16] are used to make electrical

---

[16] The pattern of a thermal relief pad is often referred to as a "wagon wheel," because the links to the plated-through hole or via look like the spokes of a wheel. Depending on a number of factors, a thermal relief pad may have anywhere from one to four spokes.

**Figure 18.19: Multilayer boards:
power and ground planes, and
thermal relief and anti-pads**

connections to the power and ground planes. The spokes in the thermal relief pads are large enough to allow sufficient current to flow, but not so large that they will conduct too much heat.

Thermal relief pads are necessary to prevent excessive heat from being absorbed into the ground and power planes when the board is being soldered.[17] When the solder is applied, a surface-tension effect known as capillary action sucks it up the vias and plated through-holes. The solder must be drawn all the way through to form reliable, low-resistance connections. The amount of copper contained in the power and ground planes can cause problems because it causes them to act as thermal heat sinks. The use of thermal relief pads ensures good electrical connections, while greatly reducing heat absorption. If the thermal relief pads were not present, the power and ground planes would absorb too much heat too quickly. This would cause the solder to cool and form plugs in the vias resulting in unreliable, high-resistance connections. Additionally, in the case of wave soldering, so much heat would be absorbed by the power and ground planes that all the layers forming the board could separate in a process known as *delamination*.

A special flavor of multilayer boards known as *Padcap* (or *"Pads-Only-Outer-Layers"*) are sometimes used for high-reliability military applications. Padcap boards are distinguished by the fact that the outer surfaces of the board only carry pads, while any tracks are exclusively created on inner layers and connected to the pads by vias. Padcap technology offers a high degree of protection in hostile environments because all of the tracks are inside the board.

Discrete Wire Technology

Discrete wire technology is an interesting discipline that has enjoyed only limited recognition, but may be poised to emerge as the technology-of-choice for designs which demand the highest signal speeds and component densities. Circuit boards created using this technology are known as discrete wired boards (DWBs).

---

[17] For future reference, the term *pad-stack* refers to any pads, anti-pads, and thermal relief pads associated with a particular via or plated-through hole as it passes through the board.

The earliest form of discrete wire technology, commonly known as *multiwire*,[18,19] was developed in the late 1960s, and had gained both military and commercial acceptance by the late 1970s. The discrete wire process commences with a conventional FR4 base layer with copper foil bonded to both sides to form the power and ground planes.[20] After any thermal relief pads and anti-pads have been etched into the copper using conventional opto-lithographic processes (Figure 8.20), a layer of insulating prepreg is bonded to each side and cured. This is followed by a layer of adhesive,[21] or *wiring film*, which is applied using a hot roll laminating machine.

A special computer-controlled wiring machine[22] is used to ultra-sonically bond extremely fine insulated wires[23] into the wiring film. The wire is routed through all the points to which it will eventually be connected. When the last point of a particular net is reached, the wiring machine cuts the wire and positions the wiring head at the next starting point (Figure 8.21).

The wire has an insulating coat of polyimide and can be wired orthog-onally (north-south and east-west), diagonally, or as a combination of both. Due to the fact that the wires are insulated, they can cross over each other without making any unwanted electrical connections. After all the wires have been applied to one side of the board, the board is inverted and more wires are applied to the other side. The majority of such boards have just these two wiring layers, one on each side. However, if necessary, additional layers of prepreg and wiring film can be bonded to the outer surfaces of the board and more wiring layers can be applied. It is not unusual for a very

---

[18] Multiwire is a registered trademark of Advanced Interconnection Technology (AIT), Islip, New York, USA.

[19] Many thanks for the wealth of information on discrete wire technology which was provided by Hitachi Chemical Electro-Products Division (Atlanta, Georgia, USA), I-CON Industries (Euless, Texas, USA), and MW Technologies (Aldershot, Hampshire, England).

[20] This discussion concentrates on a reasonably simple implementation, more complex boards with multiple and/or split power and ground planes are also available.

[21] Manufacturers of discrete wire boards are trying to discard the term "adhesive" on the basis that this layer is not actually "sticky."

[22] This is usually referred to as an NC wiring machine, where NC stands for "Numerically Controlled."

[23] The wires used in the original process were 0.16 mm in diameter. Today's processes use additional wire diameters of 0.10 mm and 0.06 mm.

(Only the power plane is shown for clarity. The ground plane would be bonded to the bottom of the board.)

Figure 18.20: Multiwire boards: preparing the core

Figure 18.21: Multiwire boards: ultrasonically bonding the wire

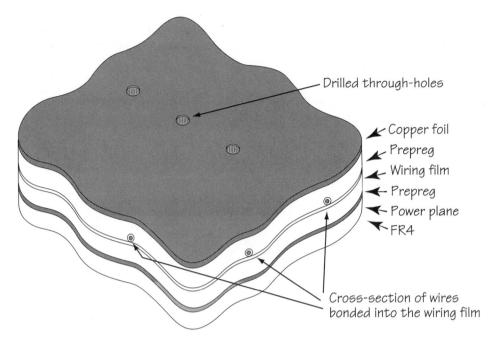

Drilled through-holes

Copper foil
Prepreg
Wiring film
Prepreg
Power plane
FR4

Cross-section of wires
bonded into the wiring film

**Figure 18.22: Multiwire boards:
drilling the holes**

Fan-out vias and pads

Component pads
(surface mount devices)

Hole drilled for via
cuts through wire.
Via's plating con-
nects to the wire

Thermal relief pad
connects this via's plating
into the power plane

Anti-pad prevents this via's
plating from connecting
into the power plane

**Figure 18.23: Multiwire boards:
plating and etching**

dense multiwire board to support four wiring layers, two on each side. Proponents of this technology say that an equivalent multilayer board might require twenty or more signal layers to achieve the same effect.

After all of the wiring layers have been created, a final layer of prepreg and copper foil are laminated onto the outer surfaces of each side of the board, and any necessary holes are drilled through the board (Figure 18.22).

The drilling process leaves the wire ends exposed in the holes. The board is now exposed to a polyimide-selective etchant which etches away the insulation at the end of wires. Although the insulation is only etched away to a depth of approximately 0.05 mm, this is sufficient for the subsequent plating process to form a *wrap-around joint,* which provides mechanical reliability as opposed to a simple electrical connection.

The holes are plated with copper to form plated through-holes and vias, and then the outer surfaces of the board are etched to form any via pads and component mounting pads (Figure 18.23). Note that the pads in this illustration indicate that the board is to be populated with surface mount devices. Finally, the board is tinned, solder masks are applied, and components are attached in the same way as for standard printed circuit boards. In fact, a completed multiwire board appears identical to a printed circuit board.

If multiwire is a niche market, then its younger brother, *microwire,*[24] forms a niche within a niche. Microwire augments the main attributes of multiwire with laser-drilled blind vias, allowing these boards to support the maximum number of wires and components.[25] Due to the large numbers of component they support, microwire boards typically have to handle and dissipate a great deal of heat, so their cores may be formed from sandwiches of copper-Invar-copper[26] or similar materials. This sandwich structure is used because the resulting *coefficient of thermal expansion* of the core combined with the other materials used in the board is almost equal to that

---

[24] Microwire is a registered trademark of Advanced Interconnection Technology (AIT), Islip, New York, USA.

[25] Actually, as multiwire boards can also be created with blind vias, and as the diameter of these vias can be as small as their microwire equivalents (around 0.25 mm), someone less trusting than the author might wonder whether the main justification for microwire was the fact that someone simply wanted to play with lasers.

[26] Invar is an alloy similar to bronze.

of any components with ceramic packages that may be attached to the board. The coefficient of thermal expansion defines the amount a material expands and contracts due to changes in temperature. If materials with different coefficients are bonded together, changes in temperature will cause shear forces at the interface between them. Engineering the board to have a similar coefficient to any ceramic component packages helps to ensure that a change in temperature will not result in those components leaping off the board.

To provide a sense of scale, in one of the more common microwire implementations each copper-Invar-copper sandwich is 0.15 mm thick. The board's central core is formed from two of these sandwiches separated by a layer of insulating prepreg, where the upper sandwich forms the power plane and the lower sandwich forms the ground plane[27] (Figure 18.24).

Thermal relief pads and anti-pads are etched into the upper and lower copper-Invar-copper sandwiches in the same way as for multiwire boards. The etch is performed through all of the layers forming the sandwich, all the way down to the surface of the prepreg in the center. In addition to the standard thermal relief and anti-pads, microwire boards also contain special pads known as *bounce pads* for use during the laser drilling process discussed below.

Following the etching process, a layer of liquid epoxy is applied to the outer surfaces.[28] After this epoxy has been cured, a layer of wiring film is applied into which wires are ultrasonically bonded as for multiwire boards. However, in the case of microwire, the wires are always 0.06 mm in diameter (Figure 18.25).

As with multiwire, the majority of microwire boards have just two wiring layers, one on each side. However, if necessary, additional layers of liquid epoxy and wiring film can be bonded to the outer surfaces of the board and more wiring layers can be applied.

---

[27] More complex structures are also possible, including multiple and/or split power and ground planes.

[28] Liquid epoxy is used to cover the wires because the glass fibers in prepreg would be impervious to the subsequent laser-drilling operations.

(Only the power plane is shown
for clarity. The ground plane
would be bonded to the bottom
of the fiberglass substrate.)

**Figure 18.24: Microwire boards:
preparing the core**

**Figure 18.25: Microwire boards:
ultrasonically bonding the wire**

Mechanically drilled
through-holes

Laser-drilled blind hole.
The laser beam strips the
polyimide insulation off the wire.

← Cured epoxy
← Wiring film
← Cured epoxy
← Power plane
← FR4

Bounce pad
reflects laser beam

**Figure 18.26: Microwire boards:
laser-drilling the blind vias**

Anti-pad prevents this
through-hole via's plating
from connecting into the
power plane

Component pads plated
onto surface of board

Plating the vias connects
the pads to the wires

Thermal relief pad
connects this through-hole via's
plating into the power plane

Ring etched around the bounce pad
acts like an anti-pad and prevents
this blind via's plating from
connecting into the power plane

**Figure 18.27: Microwire boards:
plating the vias and pads**

After all of the wiring layers have been created, a final encapsulation layer of liquid epoxy is applied to the outer surfaces of the board and cured. Now comes the fun part of the process, in which blind vias are drilled using a carbon-dioxide laser (Figure 18.26).

The laser evaporates the organic materials and strips the polyimide coating off the top of wire, but does not affect the wire itself. The laser beam continues down to the bounce pad which reflects, or bounces, it back up, thereby stripping the polyimide coating off the bottom of the wire. To complete the board, holes for through-hole vias and, possibly, plated through-holes are created using a conventional drilling process. The through-hole vias, laser-drilled blind vias, and plated through-holes are plated with copper, along with any via pads and component pads which are plated onto the board's outer surfaces (Figure 18.27), and standard tinning and solder mask processes are then applied.

Since its inception, discrete wire technology has attracted a small number of dedicated users, but has never achieved widespread recognition. Both multiwire and microwire processes are more expensive than their conventional printed wire counterparts, and therefore they are normally reserved for the most dense and electrically sophisticated applications. In fact, this discipline was originally regarded by many designers as simply being a useful prototyping technology which offered fast turnaround times on design modifications. However, as many of its devotees have known for a long time, discrete wire technology offers a number of advantages over traditional printed wire boards. One obvious advantage is that discrete wired boards with only two wiring layers (one on each side) can provide an equivalent capability to ten or more signal layers in their multilayer counterparts. Thus, discrete wire technology can offer substantially thinner and lighter circuit boards.

Additionally, discrete wire technology is particularly advantageous in the case of high-speed applications. One major requirement of high-speed designs is to control the impedance (capacitance and inductance) of the interconnect, where the impedance is a function of the distance of the tracks from the power and ground planes. In the case of multilayer boards, the tracks keep on changing their distance from the planes as they pass through vias from one layer to another. To compensate for this, additional ground planes must be added between each pair of signal layers, which increases the thickness, weight, and complexity of the boards. By comparison, the wires on discrete wire boards maintain a constant distance from the planes.

Another consideration is the vias themselves. Transmitting a signal down a conducting path may be compared to shouting down a corridor, where any sharp turns in the corridor cause reflections and echoes. Similarly, the vias in a multilayer board have significant capacitive and inductive parasitic effects associated with them, and these effects cause deterioration and corruption of the signals being passed through them. In the case of discrete wire technology, the points at which wires cross over each other have similar effects, but they are orders of magnitude smaller.

Yet another consideration is that, as the frequency of signals increases, electrons start to only be conducted on the outer surface, or skin, of a con-ductor. This phenomenon is known as the *skin effect*. By the nature of their construction, the etched tracks on multilayer boards have small imperfections at their edges. These uneven surfaces slow the propagation of signals and cause noise. By comparison, the wire used in discrete wire technology is uniform to ± 0.0025 mm, which is smoother than etched tracks by a factor of 10:1.

Last but not least, the optimal cross-section for a high-speed conductor is circular rather than square or rectangular, and a constant cross-section should be maintained throughout the conductor's length. Both of these criteria are met by discrete wire technology which, in fact, offers interconnections that are close to ideal transmission lines. The end result is that discrete wire boards can operate at frequencies that are simply not possible with their traditional multi-layer counterparts.[29] It is for this reason that discrete wire technology may be poised to emerge as the technology-of-choice for high-speed designers.[30]

---

[29] The rumor on the streets is that discrete wire circuit boards can operate at frequencies up to 600 megahertz without employing any particularly restrictive design rules, and above 1.0 gigahertz with the aid of appropriate design rules. Additionally, discrete wire boards can be constructed using microcoaxial cable, and such boards can operate at frequencies that will make your eyes water!

[30] If you are interested in obtaining more information on discrete wire technology, including a worldwide list of manufacturers and qualified design bureaus, you can contact Advanced Interconnection Technology (AIT), Islip, New York, USA.

Backplanes and Motherboards

Yet another flavor of multilayer boards, known as *backplanes*, have their own unique design constraints and form a subject in their own right. Backplanes usually have a number of connectors into which standard circuit boards are plugged (Figure 18.28).

Backplanes typically do not carry any active components such as integrated circuits, but they often carry passive components such as resistors and capacitors. If a backplane does contain active components, then it is usually referred to as a *mother board*. In this case, the boards plugged into it are referred to as *daughter boards*.

Because of the weight that they have to support, backplanes for large systems can be 1 cm thick or more. Their conducting layers have thicker copper plating than standard boards and the spaces between adjacent tracks are wider to reduce noise caused by inductive effects. Backplanes may also support a lot of hardware in the form of bolts, earth straps, and power cables. It is not unusual for a backplane to have multiple power planes such as +5 volts, −5 volts, +12 volts, −12 volts, +24 volts, and −24 volts. Each power plane typically has an independent ground plane associated with it to increase noise immunity. So, our hypothetical backplane would require 12 conducting layers for the power and ground planes alone. Additionally, systems

Figure 18.28: Backplanes

containing both analog and digital circuits often require independent power and ground planes for purposes of noise immunity. Backplanes also require excellent thermal tolerance, because some of the more heroic systems can consume upwards of 200 amps and generate more heat than a rampaging herd of large electric radiators.

Backplanes may be constructed using multilayer techniques or discrete wire technology. In fact, discrete wire technology is starting to be used big time in backplanes because of its high performance, and also due to the resulting reduction in layers, thickness, and weight.

Conductive Ink Technology

The underlying concept of conductive ink technology is relatively simple. Tracks are screen printed onto a bare board using a conducting ink, which is then cured in an oven. Next, a dielectric, or insulating, layer is screen printed over the top of the tracks. The screen used to print the dielectric layer is patterned so as to leave holes over selected pads on the signal layer. After the dielectric layer has been cured, the cycle is repeated to build a number of signal layers separated by dielectric layers. The holes patterned into the dielectric layers are used to form vias between the signal layers:

Figure 18.29: Conductive ink technology

Finally, plated through-holes and vias can be created, and components can be mounted, using the standard processes described previously. The apparent simplicity of the conductive ink technique hides an underlying sophistication in materials technology. Early inks were formed from resin pastes loaded with silver or copper powder. These inks required high firing temperatures to boil off the paste and melt the powder to form conducting tracks. Additionally, the end product was not comparable to copper foil for adhesion, conductivity, or solderability.

In the early 1990s, new inks were developed based on pastes containing a mixture of two metal or alloy powders. One powder has a relatively low melting point, while the other has a relatively high melting point. When the board is cured in a reflow oven at temperatures as low as 200°C, a process called *sintering* occurs between the two powders, resulting in an alloy with a high melting point and good conductivity.

Conductive ink technology has not yet achieved track widths as fine as traditional circuit board processes, but it does have a number of attractive features, not the least that it uses commonly available screen-printing equipment. Modern inks have electrical conductivity comparable to copper and they work well with both wave soldering and reflow soldering techniques. Additionally, these processes generate less waste and are more cost-effective and efficient than the plating and etching of copper tracks.

Chip-On-Board (COB)

Chip-on-board (COB) is a fairly modern process that only began to gain widespread recognition in the early 1990s. As the name implies, unpackaged integrated circuits are mounted directly onto the surface of the board. The integrated circuits are mechanically and electrically connected using similar techniques to those used for hybrids and multichip modules.[31] The final step is encapsulation, where the integrated circuits and their connections are covered with "globs" of epoxy resin or plastic, which are then cured to form hard protective covers[32] (Figure 18.30).

---

[31] Hybrids and multichip modules are introduced in Chapters 19 and 20, respectively.

[32] In addition to mechanical and environmental protection, the encapsulating material is also used to block out light.

Figure 18.30: Chip-on-board (COB)

There are a number of variations of chip-on-board. For example, the designer may wish to maintain an extremely low profile for applications such as intelligent credit cards. One way to achieve this is to form cavities in the board into which the integrated circuits are inserted. Compared to surface mount technology, and especially to through-hole technology, chip-on-board offers significant reductions in size, area, and weight. Additionally, this technique boosts performance because the chips can be mounted closer together, resulting in shorter tracks and faster signals.

Flexible Printed Circuits (FPCs)

Last, but not least, are flexible printed circuits (FPCs), often abbreviated to *flex*, in which patterns of conducting tracks are printed onto flexible materials. Surprisingly, flexible circuits are not a recent innovation—as they can trace their ancestry back to 1904 when conductive inks were printed on linen paper. However, modern flexible circuits are predominantly made from organic materials such as polyesters and polyimides. These base layers can be thinner than a human hair, yet still withstand temperatures up to approximately 700°C without decomposing.

There are many variants of flexible circuits, not the least being flexing, or *dynamic flex*, and non-flexing, or *static flex*. Dynamic flex is used in applications that are required to undergo constant flexing such as ribbon cables in printers, while static flex can be manipulated into permanent three-dimensional shapes for applications such as calculators and high-tech cameras requiring efficient use of volume and not just area:

Flexible circuit manipulated into a three-dimensional shape

Integrated circuits

Figure 18.31: Flexible printed circuits: static flex

As well as single-sided flex, there are also double-sided and multilayer variants. Additionally, unpackaged integrated circuits can be mounted directly onto the surface of the flexible circuits in a similar manner to chip-on-board discussed above. However, in this case, the process is referred to as chip-on-flex (COF). A common manifestation of flex technology is found in hybrid constructions known as *rigid flex,* which combine standard rigid circuit boards with flexible printed circuits (Figure 18.32).

In this example, the flexible printed circuit linking the rigid boards eliminates the need for connectors on the boards which would have to be linked by cables, thereby reducing the component count, weight, and susceptibility to vibration of the circuit, and greatly increasing its reliability.

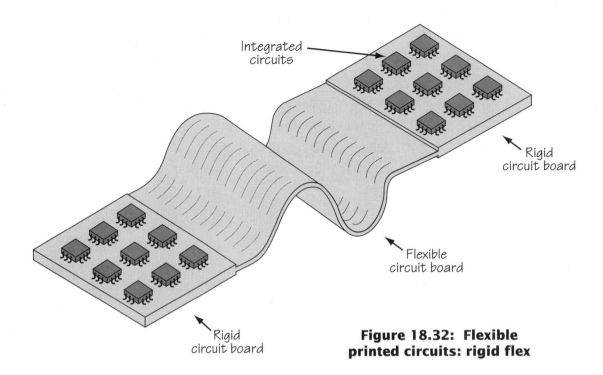

Integrated circuits

Rigid circuit board

Flexible circuit board

Rigid circuit board

Figure 18.32: Flexible printed circuits: rigid flex

While the use of flexible circuits is relatively low, it is beginning to increase for a number of reasons. These include the ongoing development of miniaturized, light-weight, portable electronic systems such as cellular phones, and the maturing of surface mount technology, which has been described as the ideal packaging technology for flexible circuits. Additionally, flexible circuits are amenable to being produced in the form of a continuous roll, which can offer significant manufacturing advantages for large production runs.

Hybrids

The word hybrid is defined as *"the offspring resulting from crossbreeding."* Many would agree that this is an apt description for the species of electronic entities known as *hybrids*, which combine esoteric mixtures of interconnection and packaging technologies. In electronic terms, a hybrid consists of a collection of components mounted on a single insulating base layer called the *substrate*. A typical hybrid may contain a number of packaged or unpackaged integrated circuits and a variety of discrete components such as resistors, capacitors and inductors, all attached directly to the substrate. Connections between the components are formed on the surface of the substrate; also, some components such as resistors and inductors may be fabricated directly onto the surface of the substrate.

Hybrid Substrates

Hybrid substrates are predominantly formed from alumina (aluminum oxide) or similar ceramic materials. Ceramics have many valuable properties which have been recognized since the Chinese first created their superb porcelains during the Ming dynasty. In addition to being cheap, light, rugged, and well-understood, ceramics have a variety of characteristics that make them particularly well suited to electronic applications. They are nonporous and do not absorb moisture, they can be extremely tough,[1] they have very good lateral thermal conductivity, and their coefficient of thermal expansion is close to that of silicon. Good lateral thermal conductivity means that heat generated by the components can be conducted horizontally across the substrate and out through its leads.

---

[1] For example, ceramics can be used to create artificial bone joints or to line the faces of golf clubs.

The coefficient of thermal expansion defines the amount a material expands and contracts due to changes in temperature. If materials with different coefficients are bonded together, changes in temperature will cause shear forces at the interface between them. Because silicon and ceramic have similar coefficients of thermal expansion, they expand and contract at the same rate. This is particularly relevant when unpackaged silicon chips are bonded directly to the hybrid's substrate, because it helps to ensure that a change in temperature will not result in the chips leaping off the substrate.

Hybrid substrates are usually created by placing ceramic powder in a mold and firing it at high temperatures. The resulting substrates have very smooth surfaces and are flat without any significant curvature. Hybrid substrates typically range in size from 2.5 cm × 2.5 cm to 10 cm × 15 cm and are in the order of 0.8 mm thick.

Some hybrids require numbers of small holes called *vias* between the top and bottom surfaces of the substrate. These vias will eventually be plated to form conducting paths between the two surfaces. Additionally, depending on the packaging technology being used, slightly larger holes may be required to accommodate the hybrid's leads. One technique for forming these vias and lead-holes is to introduce tiny pillar structures into the mold. The ceramic powder flows around the pillars and, when the mold (including the pillars) is removed after firing, the substrate contains holes corresponding to the pillars. Until recent times there was no cost-effective method for drilling vias through a ceramic substrate after it had been fired.[2] However, developments in laser technology have made it possible to punch holes through fired substrates using laser beams.

While the majority of hybrid substrates are ceramic, a wide variety of other materials may also be employed. These include glass, small FR4 circuit boards (laminated substrates), and even cardboard. The latter may have appeared among your Christmas presents embedded in a pair of socks that play an annoying tune when squeezed.[3]

---

[2] A process not dissimilar to attempting to bore holes through a dinner plate.

[3] Thank you, Auntie Barbara, I wear them all the time.

The Thick-Film Process

The two most common techniques used to create tracks and components on the surface of hybrid substrates are known as the *thick-film* and *thin-film* processes. The thick-film process is based on screen printing, an ancient art whose invention is usually attributed to the Chinese around 3,000 BC.

An optical mask is created carrying a pattern formed by areas that are either transparent or opaque to ultraviolet frequencies. The simple patterns shown in the following diagrams are used for reasons of clarity; in practice, such a mask may contain hundreds or thousands of fine lines and geometric shapes. Next, an extremely fine steel mesh the same size as the hybrid substrate is coated with a layer of photo-resistive emulsion:

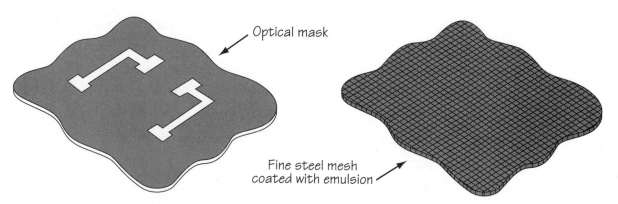

Figure 19.1: Thick-film optical mask and emulsion-coated fine steel mesh

The emulsion-coated mesh is first dried, then baked, and then exposed to ultraviolet radiation passed through the optical mask. The ionizing radiation passes through the transparent areas of the optical mask into the emulsion where it breaks down the molecular structure of the resist. The mesh is then bathed in a solvent which dissolves the degraded resist; thus, the pattern on the optical mask has been transferred to a corresponding pattern in the resist. The steel mesh with the patterned resist forms a screen-print mask, through which a paste containing metal and glass particles suspended in a solvent is applied onto the surface of the substrate (Figure 19.2).

**Figure 19.2: Thick-film tracks
screen-printed onto hybrid substrate**

The metal particles suspended in the paste are usually those of a noble metal such as gold, silver, or platinum, or an alloy of such metals such as pladium-silver (platinum and silver). When the substrate is dried, the solvent evaporates, leaving the particles of glass and metal forming tracks on the surface of the substrate. Thick-film tracks are in the order of 0.01 mm thick. The widths of the tracks and the spaces between adjacent tracks are normally in the order of 0.25 mm, but can be as low as 0.1 mm in a leading-edge process.

Multiple layers of tracks can be printed onto the surface, each requiring the creation of a unique screen-print mask. A pattern of insulating material called a *dielectric layer* must be inserted between each pair of tracking layers to keep them separated. The dielectric patterns are formed from a paste containing only glass particles suspended in a solvent. Each dielectric layer requires its own screen-print mask and is applied using an identical process to that for the tracking layers. Holes are included in the dielectric patterns where it is required for tracks from adjacent tracking layers to be connected to each other. Typical hybrids employ four tracking layers, commercial applications are usually limited to between seven and nine tracking layers, and a practical limit for current process technologies is around 14 tracking layers.[4]

---

[4] The surface of the substrate becomes increasingly irregular with every layer that is applied. Eventually, the screen-print mask does not make sufficiently good contact across the substrate's surface and paste "leaks out" under the edges of the patterns.

More layers can be used, but there is a crossover point where falling yields make the addition of successive layers cost-prohibitive. When all the tracking and insulating layers have been laid down and dried, the substrate is re-fired to approximately 1,000°C.

Resistors can be formed from a paste containing carbon compounds suspended in a solvent; the mixture of carbon compounds determines the resistivity of the final component. The resistors require their own screen-print mask and are applied to the substrate using an identical process to that described above.

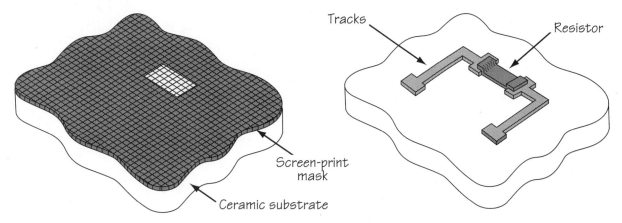

Figure 19.3: Thick-film resistors screen-printed onto substrate

When the substrate is dried, the solvent evaporates, leaving the carbon compounds to form resistors on the surface of the substrate. Assuming a constant thickness, each resistor has a resistance defined by its length divided by its width and multiplied by the resistivity of the paste. Thus, multiple resistors with different values can be created in a single screen-print operation by controlling the length and width of each component. However, if a low resistivity paste is used, resistors with large values will occupy too great an area. Similarly, if a high resistivity paste is used, resistors with small values will be difficult to achieve within the required tolerances. To overcome these limitations, the process may be repeated with a series of screen-print masks combined with pastes of different resistivity.

When all of the resistors have been created, the substrate is refired to approximately 600°C. An additional screen-print operation is employed to lay a protective overglaze over the resistors. This protective layer is formed from a paste containing glass particles in a solvent and is fired at approximately 450°C.

Unfortunately, creating resistors as described above is not as exact a process as one could wish. To compensate for process tolerances and achieve precise values, the resistors have to be trimmed using a laser beam. There are two types of laser trimming, known as *passive trimming* and *active trimming*. Passive trimming is performed before any integrated circuits or discrete components are mounted on the substrate. Probes are placed at each end of a resistor to monitor its value, while a laser beam is used to cut parts of the resistive material away. There are a variety of different cuts which may be used to modify the resistor, including plunge cuts, double plunge cuts, or L-shaped cuts.

(a) Plunge cut   (b) Double plunge cut   (c) L-shaped cut

Figure 19.4: Thick-film laser trimming resistors

The L-shaped cut combines a plunge cut with a second cut at 90°; the plunge cut provides a coarse alteration and the second cut supplies finer modifications. After each resistor has been trimmed, the probes are automatically moved to the next resistor and the process is repeated.

By comparison, active trimming is used to fine-tune analog circuits (such as active filters) and requires the integrated circuits and discrete components to be mounted on the substrate. The whole circuit is powered up and the relevant portion of the circuit is stimulated with suitable signals. A probe is placed at the output of the circuit to monitor characteristics such as amplification and frequency response. A laser is used to trim appropriate resistors to achieve the required characteristics while the output of the circuit is being monitored.

Capacitors and inductors can also be fabricated directly onto the substrate. However, capacitors created in this way are usually not very accurate and discrete components are typically used. If inductors are included on the substrate, they are created at the same time and using the same paste as one of the tracking layers. There are two main variations of such inductors: spiral and square spiral.

(a) Spiral (b) Square spiral

Figure 19.5: Thick-film inductors screen-printed onto substrate

The connection to the center-tap of the inductor can be made in several ways. A wire link can be connected to the center-tap, arched over the paths forming the spiral, and connected to a pad on the substrate outside the spiral. A somewhat similar solution is to use a track on another tracking layer to connect the center-tap to a point outside the spiral. As usual, an insulating dielectric layer would be used to separate the layer forming the inductor from the tracking layer. In yet another alternative, the center-tap can be connected to tracks on the bottom side of the substrate by means of a via placed at the center of the spiral.

Thick film hybrids can support tracking layers on both sides of the substrate, with vias being used to make any necessary connections between the two sides. Components of all types can be mounted on both sides of the substrate as required, but active components such as integrated circuits are usually only mounted on the upper side.

The Thin-Film Process

Thin-film processes typically employ either ceramic or glass substrates. The substrate is prepared by spluttering a layer of nichrome (nickel and chromium) alloy across the whole of its upper surface, then electroplating a layer of gold on top of the nichrome. The nichrome sticks to the substrate and the gold sticks to the nichrome. The nichrome and gold layers are each in the order of 5 μm (five-millionths of a meter) thick.

The thin-film process is similar to the opto-lithographic processes used to create integrated circuits. An optical mask is created carrying a pattern formed by areas that are either transparent or opaque to ultraviolet frequencies. As usual, the simple patterns shown in the following diagrams are used for reasons of clarity; in practice, such a mask may contain hundreds of thousands of fine lines and geometric shapes.

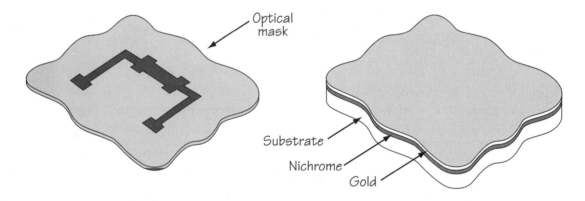

Figure 19.6: Thin-film optical mask and substrate

The surface of the gold is coated with a layer of photo-resistive emulsion, which is first dried then baked, and then exposed to ultraviolet radiation passed through the optical mask. The ionizing radiation passes through the transparent areas of the mask to break down the molecular structure of the emulsion. The substrate is bathed in a solvent which dissolves the degraded resist, then etched with a solvent that dissolves both the gold and nichrome from any areas left unprotected. The nichrome and gold remaining after the first mask-and-etch sequence represent a combination of tracks and resistors (Figure 19.7).

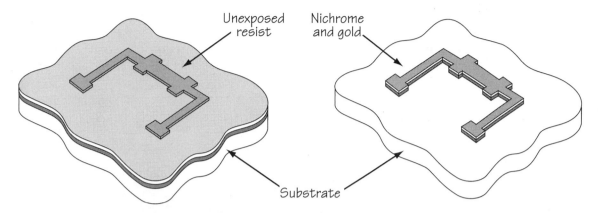

Figure 19.7: Thin-film combined tracks and resistors

The thin-film tracks are typically in the order of 0.025 mm in width, but can be as narrow as 0.001 mm in a leading-edge process. The substrate is recoated with a second layer of emulsion and the process is repeated with a different mask. The solvent used in this iteration only dissolves any exposed gold, but does not affect the underlying nichrome. The gold is removed from specific sites to expose the nichrome underneath, and it is these exposed areas of nichrome that form the resistors:

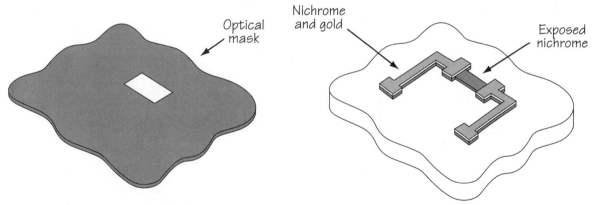

Figure 19.8: Thin-film separated tracks and resistors

In certain respects, thin-film designers have less freedom in their control of resistance values than do their thick-film counterparts. Although the resistivity of the nichrome layer can be varied to some extent from hybrid to hybrid, the resistivity for a single hybrid is constant across the whole surface. Thus, the only way to select the value of an individual resistor is by controlling its length and width. In addition to simple rectangles, thin-film resistors are often constructed in complex concertina shapes with associated trimming blocks. The resistor values can be subsequently modified by laser trimming.

(a) Before trimming (b) After trimming

Figure 19.9: Thin-film laser trimming resistors

Resistors may also be created in ladder structures and their values modified by selectively cutting out rungs from the ladder using the laser. Thin-film hybrids normally only employ one tracking layer on a single side of the substrate; however, multilayer and double-sided variations are available.

The Assembly Process

Some hybrids contain only passive components such as resistor–capacitor networks; others contain active devices such as integrated circuits. If integrated circuits are present, they may be individually packaged surface mount devices or unpackaged bare die. Individually packaged integrated circuits are mounted onto the substrate using similar techniques to those used for surface mount technology printed circuit boards.[5] Unpackaged die are mounted directly onto the substrate using the techniques discussed below.

---

[5] These techniques were introduced in Chapter 18.

The primary method of removing heat from an unpackaged integrated circuit is by conduction through the back of the die into the hybrid's substrate. One technique for attaching die to the substrate is to use an adhesive such as silver-loaded epoxy. The epoxy is screen-printed onto the sites where the die are to be located, the die are pushed down into the epoxy by an automated pick-and-place machine, and the epoxy is then cured in an oven at approximately 180°C:

Figure 19.10: Bare die attached to substrate

The diagrams shown here have been highly simplified for reasons of clarity; in practice, a hybrid may contain many die, each of which could have numerous pads for power and ground connections and hundreds of pads for input and output signals.

An alternative method of die attachment is that of a *eutectic bond*, which requires *gold flash* (a layer of gold with a thickness measured on the molecular level) on the back of the die and also at the target site on the substrate. The substrate is heated to approximately 300°C, then an automatic machine presses the die against the substrate and vibrates it at ultrasonic frequencies to create a friction weld. The process of vibrating the die ultrasonically is called *scrubbing*. Eutectic bonds are mainly used for military applications that are intended to withstand high accelerations.

After the die have been physically attached to the substrate, electrical connections are made between the pads on each die and corresponding pads on the substrate. The most commonly used technique is *wire bonding*, in which an automatic machine makes the connections using gold or aluminum wires. These bonding wires are finer than a human hair, the gold and aluminum wires typically having diameters in the order of 0.05 mm and 0.025 mm, respectively. In one technique known as a ball bond, the end of the wire is heated with a hydrogen flame until it melts and forms a ball, which is then brought into contact with the target. As an alternative to ball bonds, aluminum wires are usually attached by pressing the wire against the target and scrubbing it at ultrasonic frequencies to form a friction weld. Similar wire bonds can also be used to form links between tracks on the substrate and to form connections to the hybrid's pins:

Figure 19.11: Wire bonding

Another die attachment process is that of *tape automated bonding* (*TAB*), which is mainly used for high-volume production runs or for devices with a large number of pads. In this process, a transparent flexible tape is coated with a thin layer of metal into which lead frames are patterned using standard opto-lithographic techniques. The pads on the bare die are attached to corresponding pads on the tape, which is then stored in a reel (Figure 19.12).

In practice, the lead frame tracks on the tape can be so fine and so close together that they are difficult to distinguish with the human eye. When the time comes to attach the die to the hybrid's substrate, silver-loaded epoxy is screen-printed onto the substrate at the sites where the devices are to be located. This silver-loaded epoxy is also printed onto the substrate's pads to which the TAB leads are to be attached. The reel of

Figure 19.12: Die attached to TAB tape

TAB tape with attached die is fed through an automatic machine which pushes the die and the TAB leads down into the epoxy at their target sites. When the silver-loaded epoxy is subsequently cured, it forms electrical connections between the TAB leads and the pads on the substrate.

Figure 19.13: Tape automated bonding (TAB)

As an alternative to all of the above, there are several attachment techniques grouped under the heading of *flipped chip*. These may be used where the substrate's real estate is at a premium and the die have to be mounted very closely together. One flipped chip technique called *solder bumping* requires perfect spheres of solder to be formed on the pads of the die. The die are then "flipped over" to bring their solder bumps into contact with corresponding pads on the substrate (Figure 19.14).

**Figure 19.14: Flipped chip –
solder bumping**

One advantage of the solder bumping technique is that pads are not obliged to be at the periphery of a die; connections may be made to pads located anywhere on a die's surface. When all of the die have been mounted on the substrate, the solder bumps can be melted in a reflow oven using infrared radiation or hot air. Alternatively, the solder may be melted using *vapor-phase soldering* in which the board is lowered into the vapor-cloud above a tank containing boiling hydrocarbons. However, as for printed circuit boards, vapor-phase soldering is becoming increasingly less popular due to environmental concerns.

Another flipped chip alternative is *flipped TAB* which, like standard TAB, may be of use for high volume production runs or for devices with a large number of pads (Figure 19.15).

Hybrids in which bare die are attached directly to the substrate are smaller, lighter, and inherently more reliable than circuits using individually packaged integrated circuits. Every solder joint and other form of connection in an assembly is a potential failure mechanism. Individually packaged integrated circuits already have a connection from each pad on the die to the corresponding package lead and require a second connection from this lead to the substrate. By comparison, in the case of a bare die, only a single connection is required between each pad on the die and its corresponding pad on the substrate. Thus, using bare die results in one less level of interconnect, which can be extremely significant when using die with hundreds of pads.

**Figure 19.15: Flipped chip –
flipped TAB**

The Packaging Process

Hybrids come in a tremendous variety of shapes, sizes, and packages. Many hybrids are not packaged at all, but simply have their external leads attached directly to the substrate. In this case, any bare die are protected by covering them with blobs of plastic polymer in a solvent; the solvent is evaporated, leaving the plastic to protect the devices:[6]

Figure 19.16: Bare die encapsulated in plastic

---

[6] This is very similar to the chip-on-board techniques introduced in Chapter 18. In fact, if the hybrid's substrate is a small printed circuit, this would fall into the category of chip-on-board.

If the hybrid is targeted for a harsh environment, the entire substrate may be coated with polymer. If necessary, additional physical protection can be provided by enclosing the substrate with a ceramic cap filled with a dry nitrogen atmosphere:

Figure 19.17: Ceramic cap package

In this case, any bare die may remain uncovered because the dry nitrogen environment protects them. If extreme protection is required, the hybrid may be hermetically sealed in alloy cans filled with dry nitrogen.

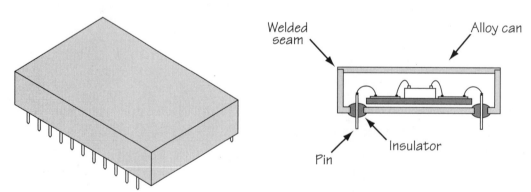

Figure 19.18: Hermetically sealed package

Finally, it should be noted that the majority of the assembly and packaging techniques introduced above are not restricted to hybrids, but may also be applied to individual integrated circuits and multichip modules. (Multichip modules are introduced in the following chapter.)

Multichip Modules (MCMs)

Traditionally, integrated circuits, hybrids, and printed circuit boards were differentiated on the basis of their substrate materials: semiconductors (mainly silicon), ceramics, and organics, respectively. Unfortunately, even this simple categorization is less than perfect. Some hybrids use laminate substrates which are, to all intents and purposes, small printed circuit boards. So, at what point does a printed circuit board become a hybrid with a laminate substrate?

The problem is exacerbated in the case of multichip modules— a generic name for a group of advanced interconnection and packaging technologies featuring unpackaged integrated circuits[1] mounted directly onto a common substrate (Figure 20.1).

Unpackaged integrated circuits

Substrate material

Figure 20.1: Multichip module

Multichip modules and hybrids can sometimes be differentiated by their choice of substrate materials and manufacturing processes. But, multichip modules can be based on almost any substrate or even combinations of substrates, including laminates and ceramics which are also used for hybrids. A slightly more reliable differentiation is that multichip modules usually have finer tracks and higher tracking and component densities[2] than hybrids.

---

[1] Including, but not limited to, analog, digital, application-specific, simple logic, memory devices, processor cores, and communications functions.

[2] In this context, the phrase "component density" does not necessarily refer to the number of integrated circuits mounted on the multichip module, but rather to the number of logic gates they represent. For example, a multichip module may consist of only four integrated circuits, but each of these could contain hundreds of thousands of logic gates. See also the discussions on "Equivalent Integrated Circuits" later in this chapter.

Even so, the classification of some devices is open for debate. In fact, many "hybrid designers" in the late 1980s were actually designing multichip modules without knowing it. This is because the term multichip module and the first definitions of what constituted these devices were only introduced in 1990.

Categorization by Substrate

The original multichip module definitions were based on categorization by their substrate material:

MCM-L Laminates such as small, fine-line printed circuit boards with copper tracks and copper vias.

MCM-C Ceramic substrates, some of which are similar to those used for hybrids, formed from single, seamless pieces of ceramic, and carrying tracks that are created using thick-film or thin-film processes (or a mixture of both). However, a large proportion of ceramic substrates for multichip modules are of the cofired variety, which is introduced in more detail below.

MCM-D Ceramic, glass, or metal substrates that are covered with a layer of dielectric material such as polyimide. The dielectric coat is used to modify the substrate's capacitive characteristics and tracks are created on the surface of the dielectric using thin-film processes.

MCM-S Semiconductor substrates, predominantly silicon, with very fine tracks formed using opto-lithographic processes similar to those used for integrated circuits. Semiconductor substrates are also known as *active substrates* because components such as transistors and logic gates can be fabricated directly onto their surface.[3] One additional benefit of using silicon as a substrate is that its coefficient of thermal expansion exactly matches that of any silicon chips that are attached to it.

---

[3] These logic gates are particularly useful for constructing special circuits to test the multichip module after it has been integrated into a larger system.

Why Use Multichip Modules?

One obvious question that may arise at this point is why use multichip modules at all? If the only requirement is to increase tracking and component densities, why not simply build bigger and better integrated circuits? After all, integrated circuit fabrication processes offer the smallest components and the finest tracks known to man. The first answer to this question is based on yield. The processes used to manufacture semiconductors are as near-perfect as modern technology allows, but they still leave a small number of molecular-level defects called *inclusions* in the crystalline structure. The presence of an inclusion will affect the manufacturing process and prevent an integrated circuit from functioning as planned.

For example, consider a semiconductor wafer that is to be used to build 10,000 integrated circuits, each 1 mm × 1 mm. Assuming that this wafer has 10 inclusions spread evenly across its surface, 10 of the chips would fail, 9,990 would work,[4] and the resulting yield would be 9,990/10,000 = 99.9%. Now consider the effect if the same wafer were used to build 100 integrated circuits, each 10 mm × 10 mm. Assuming that each inclusion affects a separate circuit, 10 of the chips would fail, 90 would work, and the yield would be 90.0%. Continuing this progression, if the wafer were used to build 25 circuits, each 20 mm × 20 mm, the yield would fall to only 60.0%.[5] Further increasing the size of the integrated circuits would rapidly reduce the yield to zero.

In addition to inclusions, problems can also arise during the doping and metallization stages—the larger the area, the more failure prone is the process. Thus, the semiconductor game is a case of the smaller the chips, the greater the chance for successful yields. The solution is to construct a number of smaller integrated circuits and package them as a single multichip module.

Another consideration is that it may be required to implement a portion of the design using a high-speed technology such as emitter-coupled logic (ECL) which uses a lot of power. If all of the logic for a multimillion gate

---

[4] This is obviously an ideal case. In fact, there are a great number of other manufacturing considerations that will also reduce the yield.

[5] This value is also based on the assumption that each inclusion would affect a separate circuit.

design were built in this technology, enough heat could be released to melt the device into a puddle. The answer is to put the speed-critical parts of the design on high-speed, high-power chips, put the less critical functions on slower, lower-power chips, and package all of the chips as a single multichip module.

However, there still remains one unanswered question—why not package each chip individually and mount them on a standard printed circuit board? The answer here is one of scale. On-chip tracks are incredibly fine compared to bonding wires, package leads, and circuit board tracks. Similarly, the lengths of on-chip tracks are measured in millimeters, unlike their circuit board equivalents, which are measured in tens of centimeters. These differences in scale have a number of consequences:

a)   The bonding wires combined with the package leads present very significant *parasitic effects* in terms of capacitance and induc-tance, which means that the logic gates driving off-chip have to handle relatively large currents. This means that the logic gates driving off-chip must be made bigger, which, in turn, means that they eat into the available silicon real estate, consume more power, and generate more heat.

b)   The longer a track is, the more susceptible it is to picking up *noise*. The main technique to overcome noise is to reduce its significance by increasing the voltage differential between the logic 0 and logic 1 values. Because on-chip tracks are so short, they are relatively immune to external noise and the voltages used to represent the logic values may differ by as little as one volt. However, circuit board tracks are much longer and are correspondingly more susceptible. Thus, the voltages used to represent the logic values at the board level typically differ by as much as five volts. The greater the difference between the voltages used to represent the logic values, the longer it takes for a signal to transition between values and the slower the speed of the system.

c) Last, but not least, signals propagate through track with a finite speed—the longer the track, the longer it takes for a signal to travel down it.

Multichip modules address all of these problems. Unpackaged integrated circuits can be mounted much closer together than their packaged counterparts, resulting in shorter tracks and faster signals. Also, the fabrication processes used for multichip modules result in much finer tracks than those found on printed circuit boards. The removal of the package leads and the shorter, finer chip-to-chip tracks means that the logic gates driving off-chip can be smaller and less power-hungry, thereby saving silicon real estate and generating less heat. Additionally, the shorter chip-to-chip tracks are less susceptible to noise, allowing the voltage differential between logic values to be reduced, thereby minimizing the time it takes for a signal to transition between logic values.

Yet another consideration is that every solder joint and other form of connection in an assembly is a potential failure mechanism. Individually packaged integrated circuits already have a connection from each pad on the die to its corresponding package lead, and they require a second connection from this lead to the substrate. By comparison, in the case of a bare die, only a single connection is required between each pad on the die and its corresponding pad on the substrate. Thus, using bare die results in one less level of interconnect, which can be extremely significant when using die with hundreds of pads.

Cofired Ceramics

Ceramic substrates have numerous benefits (discussed earlier in Chapter 19) but, after the ceramic has been fired, they are very difficult to machine and cannot be satisfactorily milled, drilled, or cut using the technology currently available.[6] One solution to this problem is cofired ceramics, in which a number of slices of ceramic are preformed and then fired together.

---

[6] Modern laser technology offers some success in drilling and cutting ceramics, but not much.

The process begins with unfired, or *"green,"* ceramic tape, which is rolled out into paper-thin layers. After each layer has been cut to the required size, microscopic holes are punched through it to form vias.[7] One technique used to punch these holes involves a steel die[8] studded with round or square pins:

Steel die with pins

Cut-away view to show steel pins

Steel die is pressed into green ceramic to form holes

Malleable green ceramic (paper-thin)

Figure 20.2: Cofired ceramics: punching holes to make vias

Unfortunately, although fast in operation, these dies take a long time to construct and the technique is incredibly expensive—each die may cost in the neighborhood of $10,000 and can typically only be used for a single layer of a single module. As an alternative, some manufacturers are experimenting with programmable variants consisting of a matrix of pins, each of which may be retracted or extended under computer control. Programmable die are very expensive tools to build initially, but can be quickly reprogrammed for multiple layers and multiple modules. Their main drawback is that all of the pins are on a fixed grid. Another approach that is routinely

---

[7] Cofired substrates typically use a tremendous amount of blind and buried vias.

[8] In this context, the term "die" refers to a piece of metal with a design engraved or embossed on it for stamping onto another material, upon which the design appears in relief.

used is that of a numerically (computer) controlled tool which punches the holes individually. This technique is relatively inexpensive, but requires much more time to complete the task.

After the holes have been punched, a screen-printing operation is used to backfill the holes with a conducting ink (a paste containing metallic powder). The mask used in this step is referred to as a *via-screen*. Next, tracks are screen-printed onto the surface of each layer using the same conducting ink. These processes are similar to the thick-film hybrid technique, but use much finer meshes for the screen print masks, thereby resulting in much finer lines and spaces.

Conductive ink screen-printed onto surface of green ceramic

Conductive ink fills holes in the ceramic

**Figure 20.3: Cofired ceramics:
screen-printing tracks**

After all of the layers have been prepared in this way, they are stacked up, pressed together, and prefired at a relatively low temperature to burn off the binders used in the conductive inks and any moisture in the ceramic. The substrate is then fully fired at around 1,600°C, which melts the tracks to form good conductors and vias, and transforms the individual layers into one homogeneous piece of ceramic (Figure 20.4).

Cofired ceramic substrates constructed in this way can contain an amazing number of layers. Substrates with 85 layers are not uncommon, while leading-edge processes may support 120 layers or more. To provide a sense of scale, an 85-layer substrate will be approximately 4.5 mm thick, while a 120-layer substrate will be around 6.25 mm thick. Finally, after the substrate has been fired, any tracks and pads on the surface layers are plated with a noble metal[9] such as gold.

---

[9] Noble metals are those such as gold and platinum, which are extremely inactive and are unaffected by air, heat, moisture, and most solvents.

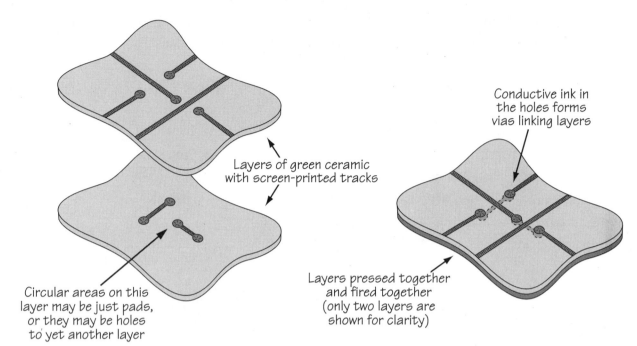

Conductive ink in the holes forms vias linking layers

Layers of green ceramic with screen-printed tracks

Circular areas on this layer may be just pads, or they may be holes to yet another layer

Layers pressed together and fired together (only two layers are shown for clarity)

Figure 20.4: Cofired ceramics: assembling and firing the layers

One not-so-obvious consideration with cofired ceramics is that the act of rolling out the ceramic tape lines up the particles of ceramic in the direction of the roll. This causes the layers to shrink asymmetrically when the ceramic is fired. To counter this effect, the layers are alternated such that adjacent layers are rolled in different directions. The net result is symmetrical shrinkage with respect to the major axes.

Low-fired Cofired Ceramics

One problem with traditional cofired processes is that the tracks must be created using *refractory metals*.[10] Although these are reasonable conductors, they are not ideal for high-reliability and high-performance applications. For high reliability one would ideally prefer a noble metal like gold, while for high performance at a reasonable price one would ideally prefer copper. (Surprisingly, copper is one of the best conductors available for

---

[10] Refractory metals are those such as tungsten, titanium, and molybdenum, which are capable of withstanding the extremely high, or refractory, temperatures necessary to fire the ceramic.

high-frequency applications, especially those operating in the microwave arena.) Unfortunately, metals like gold and copper simply vaporize at refractory temperatures.

Low-fired cofired processes use modern ceramic materials whose compositions are very different from traditional ceramics. To differentiate the two, the term *HTCC* (high-temperature cofired ceramics) is used to refer to high-temperature ceramics requiring tracks to be formed using refractory metals, while the term *LTCC* (low-temperature cofired ceramics) refers to low-fired cofired ceramics. These new materials can be fired at temperatures as low as 850°C. This allows tracks to be screen-printed using copper-based or noble metal-based conductive inks, as long as the firing takes place in a nitrogen-muffled furnace to prevent the metals from oxidizing.

Assembly and Packaging

To a large extent, multichip module packages are created on a design-specific basis. Due to the high lead counts involved, the package tends to be an integral part of the module itself. This means that packaging cannot be the afterthought it is in the case of some hybrids. For our purposes in the discussion below, we will assume a module with a cofired ceramic substrate carrying four bare die.

Because cofired substrates are constructed from multiple layers, designers have the ability to create quite sophisticated structures. For example, holes called *cavities*, or *wells*, can be cut into a number of the substrate's upper layers before they are combined together and fired (Figure 20.5).

Remembering that the layers of ceramic are paper thin, the "shelves" are actually formed from whatever number of layers are necessary to bring them to the required thickness. These wells will eventually accommodate the die, while the shelves will create a bed for the package's lid.[11] However, before the die are mounted on the substrate the external leads must be attached.

---

[11] Note that, with the exception of the upper layer, each of the layers forming the shelves will, in all probability, also be used to carry tracks.

Large holes punched in upper layers
(only two upper layers
shown for clarity)

Wells

A large number
of layers are
used to create
the full package

Cut away to show cross-
sectional view of well

All of the layers (except the top one)
may have tracks and vias
(which are not shown for clarity)

**Figure 20.5: Cofired ceramics:
creating wells**

In the case of a packaging style known as pin grid arrays (PGAs), the module's external connections are arranged as an array of conducting leads, or pins, in the base. In many PGAs the leads are flat brazed directly onto the bottom layer of the device. Alternatively, in a similar manner to the way in which the wells were created in the upper layers, smaller wells are cut into a number of the substrate's lower layers before they are combined together and fired. Each of these wells is connected to a track constructed on one of the inner layers. After firing, the wells are lined with a thin layer of metal and the leads are inserted (Figure 20.6).

The lead pitch (the spacing between adjacent leads) is typically in the order of 1.25 mm, and a module can contain more than a thousand such leads (although the average device contains rather less than this). After the leads have been inserted, the entire assembly is heated to approximately 1,000°C, which is sufficient to braze the leads to the layer of metal lining the wells. Finally, the leads are gold-plated to ensure that they will make good electrical connections when the module is eventually mounted on a circuit board.

View of bottom of pin grid array with protruding pins

Ceramic layers

Figure 20.6: Pin grid arrays

Pin brazed into well

Well formed in ceramic layers to accommodate the pin

As an alternative to pin grid arrays, some modules are presented as pad grid arrays (PGAs), in which the external connections are arranged as an array of conducting pads on the base. In yet another alternative that some observers believe is poised to become one of the premier packaging technologies of the 1990s, devices may be presented as ball grid arrays (BGAs)[12] or column grid arrays (CGAs). These forms of package are similar to a pad grid array, but with the addition of small balls or columns of solder attached to the conducting pads (Figure 20.7).

The balls (or columns) are made out of a high-temperature solder (90% tin and 10% lead) which does not melt during the soldering process. Instead, a standard low-temperature solder, referred to as eutectic solder, is used to connect the balls (or columns) to the pads. These packages are said to be self aligning because, when they are mounted on a circuit board and heated in a reflow oven, the surface tension of the eutectic solder causes the device to automatically align itself.

Ball grid arrays are physically more robust than column grid arrays. However, column grid arrays are better able to accommodate any thermal strains caused by temperature mismatches between the multichip module and the circuit board.

---

[12] Ball grid arrays may also be known as solder grid arrays, bump grid arrays, or land grid arrays.

BALL GRID ARRAY

COLUMN GRID ARRAY

Ceramic layers

Pad brazed into well

Normal solder

High-temperature solder ball (90% tin, 10% lead)

Normal solder

Copper pad on circuit board

Normal solder

High-temperature solder ball (90% tin, 10% lead)

Normal solder

Circuit board

Figure 20.7: Ball grid arrays and column grid arrays

Protective ceramic lids

Unpackaged integrated circuits

Wells are back-filled with nitrogen before the protective ceramic lids are attached

Figure 20.8: Attaching the die

In yet another variant which may be of use for specialist applications, small balls of fibrous gold known as *fuzz-buttons* are inserted between the pads on the base of the package and their corresponding pads on the board. When the package is forced against the board, the fuzz-buttons compress to form good electrical connections. Even when the pressure is removed, the fuzz-buttons act in a similar manner to Velcro™ and continue to hold the component in place. One of the main advantages of the fuzz-button approach is that it allows broken devices to be quickly removed and replaced.

However, we digress. Once the external leads have been connected, the module is ready to be populated with its die (Figure 20.8). In many ways the assembly processes used for multichip modules are similar to those used for hybrids. Bare die can be physically mounted on the substrate using adhesive, eutectic, or flipped-chip techniques, and electrically connected using wire bonding, tape automated bonding, or solder bumping methodologies. After the die have been mounted, the wells are back-filled with nitrogen and protective lids are firmly attached.

Equivalent Integrated Circuits

In previous discussions it was noted that printed circuit boards may contain thousands of components and hybrids may contain hundreds of components, but the multichip module described above contained only four integrated circuits. However, simply counting the components does not provide an appropriate basis for comparison. Although today's multichip modules typically contain relatively few devices, the actual logic represented by those devices can be huge. Thus, as opposed to *component density*, a more appropriate comparison is that of *functional density*.

There are a number of methods which may be used to compare these different interconnection strategies, including the average number of vias per square centimeter and the average number of tracks per layer per square centimeter. However, one of the most popular techniques is based on the concept of an equivalent integrated circuit.[13] The definition of an equivalent

---

[13] Similar in principle to the concept of an equivalent gate which was introduced in Chapter 14.

integrated circuit is one that contains a certain number of leads;[14] a common value is 20.

First, all of the leads associated with the real components attached to the substrate in question are added together and divided by the number of leads on the hypothetical equivalent integrated circuit. Thus, the result returned by this operation is in units of equivalent integrated circuits. For example, 300 leads divided by 20 equals 15 equivalent integrated circuits. This value is then divided by the area of the substrate to determine the number of equivalent integrated circuits per square centimeter.

To illustrate what this means in real terms, the author performed these calculations on a real printed circuit board, a real hybrid, and a real multichip module that he happened to find lying around the office. The printed circuit board was $27.5 \text{ cm} \times 20.0 \text{ cm}$ and contained 627 components with a total of 1,736 leads; the hybrid was $2.5 \text{ cm} \times 5.0 \text{ cm}$ and contained 36 components with a total of 352 leads; while the multichip module was $5.0 \text{ cm} \times 5.0 \text{ cm}$ and contained four die each with 300 leads giving a total of 1,200 leads. By some strange quirk of fate, this multichip module was the same as the one illustrated in the figures above. Performing the equivalent integrated circuit calculations returned the results shown in Figure 20.9.

Although this sample (one each of a circuit board, hybrid, and multichip module) is statistically meaningless, it does serve to illustrate the trend, which is that multichip modules offer the most sophisticated interconnection technology currently available.

	Area of substrate (cm²)	# of real component leads	# of equivalent integrated circuits	Equivalent integrated circuits per cm²
Printed circuit board	550	1,736	86.8	0.16
Hybrid	12.5	352	17.6	1.41
Multichip module	25	1,200	60	2.40

Figure 20.9: Comparison based on equivalent integrated circuits

---

[14] As usual, the actual definition of what constitutes an equivalent integrated circuit depends on whom one is talking to.

Technology Considerations

In addition to the substrates introduced above, some multichip modules use more esoteric materials such as quartz, garnet, and sapphire[15] which are targeted towards specialized applications. For example, the ultra-high frequency capabilities of sapphire are of particular interest in the microwave arena. Additionally, some multichip modules use a mixture of substrates; for example, ceramic substrates with areas of embedded or deposited silicon or diamond film. Other devices may use a mixture of tracking techniques—for example, thin-film processes for surface signals combined with thick-film processes for inner layers.

In addition to complex materials selection and process problems, designers of multichip modules are faced with a number of other challenges. Parasitic and thermal management are key issues due to the high interconnection density and high circuit speeds. Thus, it is simply not possible to split a multichip module design into an interconnection problem, a thermal management problem, and a packaging problem. Each multichip module presents a single mammoth problem in concurrent design and typically requires a unique solution. The result is that working with multichip modules can be an expensive hobby in terms of tooling requirements and long design times.

The multichip module designer's task has migrated from one having primarily electronic concerns, to one incorporating components of materials science, chemistry, and packaging. In addition to basic considerations such as selecting the packaging materials and determining the optimal connection strategy (pin grid array versus pad grid array versus ball grid array), the designer has to consider a variety of other packaging problems, including lead junction temperatures and thermal conduction and convection characteristics. The magnitude of the task is daunting, but if it was easy then everybody would be doing it, and you can hit me over the head with a kipper if it doesn't make life exciting!

---

[15] See also the discussion on diamond substrates in Chapter 21.

Alternative and Future Technologies

Electronics is one of the most exciting and innovative disciplines around, with evolutionary and revolutionary ideas appearing on almost a daily basis. Some of these ideas skulk around at the edges of the party, but never really look you in the eye or take the trouble to formally introduce themselves; some surface for a short time and then disappear forever into the twilight zone from whence they came; some tenaciously manifest themselves in mutated forms on a seasonal basis; and some leap out as if from nowhere with a fanfare of trumpets and join the mainstream so quickly that before you know it they seem like old friends.

This closing chapter introduces a smorgasbord of technologies, many of which have only recently become commercially available or are on the cutting-edge of research and development. Even the most outrageous topics presented below have undergone experimental verification, but nature is a harsh mistress and natural selection will take its toll on all but the fittest. Although some of the following may seem to be a little esoteric at first, it is important to remember that a good engineer can easily believe three impossible things before breakfast. Also remember that the naysayers proclaimed that it was impossible for bumble bees to fly (although they obviously could), that man would never reach the moon, and that I would never finish this book![1]

Reconfigurable Hardware and Interconnect

The term *"hardware"* is generally understood to refer to any of the physical portions constituting an electronic system, including components,

---

[1] Ha! I pluck my chest hairs threateningly in their general direction.

circuit boards, power supplies, cabinets, and monitors.[2] However, this discussion concentrates on the subset of hardware consisting of circuit boards, electronic components, and interconnect.

Unfortunately, the phrase *"reconfigurable hardware"* is akin to the phrase *"stretch-resistant socks"*—they both mean different things to different people. To the young and innocent, *"stretch-resistant"* would tend to imply a pair of socks that will not stretch. But, as those of us who are older, wiser, and a little sadder know, *"stretch-resistant"* actually refers to socks that *will* stretch—they just do their best to resist it for a while! Similarly, the term *"reconfigurable"* is subject to myriad diverse interpretations depending on the observer's point of view.

As a starting point, reconfigurable hardware refers to an electronic product whose function can be customized to a specific system or application. There are obvious benefits to making one product (that can be customized) many times, as opposed to making many application-specific products once. The problem is, the perception of what is implied by *"reconfigurable"* is a moving target which evolves over time as new techniques and technologies become available. Throughout most of the 1980s, the most sophisticated level of customization was displayed by products based on programmable logic devices (PLDs) such as PROMs, PLAs, and PALs,[3] or variants of read-only memory such as PROMs,[4] EPROMs, and E²PROMs.[5] Products using these devices were usually targeted at a particular application and then focused towards a specific implementation. By comparison, the advent of field-programmable gate arrays (FPGAs) in the late 1980s and early 1990s opened the door to products that could be almost totally customized for diverse applications.

---

[2] By comparison, the term *"software"* refers to intangible programs, or sequences of instructions, that are executed by hardware. Additionally, the term *"firmware"* refers to programs that are hard-coded into non-volatile memory devices, while *"vaporware"* refers to either hardware or software that exist only in the minds of the people who are trying to sell them to you.

[3] Programmable logic devices were introduced in Chapter 16.

[4] PROMs may be regarded as being programmable logic devices or as memory devices depending on the applications for which they are being used.

[5] Memory devices were introduced in Chapters 15 and 16.

As you have hopefully surmised by now, *Bebop to the Boolean Boogie* does not simply describe "where we are," but also attempts to show "how we got here." However, the pace of technological development since the birth of the integrated circuit poses particular problems in the case of reconfigurable logic, because there is no consistent terminology that has survived the ravages of time. To permit the presentation of a consistent view, the solution offered by *Bebop* is to adopt the following terminology:[6]

Configurable Hardware:	A product whose function may be customized once or a very few times.
Reconfigurable Hardware:	A product whose function may be customized many times.
Remotely Reconfigurable Hardware:	A product whose function may be customized remotely, by telephone or radio, while remaining resident in the system.
Dynamically Reconfigurable Hardware:	A product whose function may be customized "on-the-fly" while remaining resident in the system.
Virtual Hardware:	An extension of dynamically reconfigurable hardware.

And so, armed with our trusty definitions, it only remains for us to gird up our loins and delve into the ramifications of hardware whose very reason for being can be transmogrified by an idle whim.

Configurable Hardware

A classic example of a product whose function may only be customized once is provided by a car radio, of which there may be several versions: a cheap, no-frills little number for the cost-conscious buyer, a mid-range model for the young man about town, and an ultra-flash, no-holds-barred version for the powerful business executive who also delights in thrusting his

---

[6] The terminology presented here differs to some extent from current industry standards, which only tend to reflect the technology of the day (or flavor of the month).

gold Rolex under your nose. But, little do they know that it is not unusual for all of these variations to be constructed on identical circuit boards, which can be configured by adding or removing certain components and modifying certain switches or jumpers. In fact, on some occasions, the only major difference between the different models is the quality of their cases and the number of buttons and dials that they support. From the viewpoint of the manufacturer, the circuit boards used in these radios are configurable but, from the perspective of the user, their function is cast in stone.

(a) Cheap model

Figure 21.1:
Configurable hardware: car radios

(b) Expensive model

A similar example that may be a little closer to home revolves around digital wrist watches. One down-side of the electronics era is the inevitability of being cornered by someone who insists on regaling you with the 1,001 details you didn't want to know about his new digital watch that can simultaneously display the current time in Tokyo, Paris, London, New York, and Moscow, play sixteen immediately annoying tunes, and has a calculator rivaling the control panel of the space shuttle thrown in for good measure. By some strange quirk of fate, this almost invariably occurs on those days when the timepiece you are sporting arrived in the form of a free gift at the bottom of a box of cornflakes. But there is no need to lower your head in shame, because it is not beyond the bounds of possibility that both of these instruments contain identical integrated circuits! In the case of the simpler model, a hard-wired voltage level applied to one of the device's pins instructs it to pretend to be "cheap and cheerful." Once again, the major difference between the two models is the quality of their cases and . . . the price tag.

A final example of configurable products that is too good not to share revolves around a number of well-known computer manufacturers who used to offer a choice between the fast, deluxe, and expensive version of a machine, or the slower, somewhat cheaper model. However, unbeknownst to the innocent purchaser, the only difference between the two models was a simple switch on the main circuit board. Depending on the position of this switch, the system's clock either ran at full speed or was "slugged" to half its normal operating frequency. When the owners of one of the slower systems decided that they just had to upgrade, large amounts of money would change hands and, after appropriate sacrifices had been offered to the Gods, the computer engineer would (eventually) arrive. After making a ceremony of preparing to exchange the circuit boards, the engineer would suddenly leap to his feet, point excitedly out of the window, and cry "Good grief! What's that?" Then, while everybody's backs were turned, he would flick the switch and have the board halfway back into the system before anybody knew what was happening. And there are those who would say that electronics is dull and boring!

Reconfigurable Hardware

Hardware that is simply configurable is obviously limited, because everything that the product can do has to be designed into its base configuration, which has to encompass all possible variants. One technique for producing a product whose function may be extended beyond its original design objectives is to base that product on devices that can be reprogrammed. For example, a PROM could be employed in the role of a hardware truth table (Figure 21.2).

Similarly, a PLA could be used to implement a state machine that performs a certain sequence of operations, or, based on the inputs presented to it, a PAL device could be used to implement Boolean equations and to generate appropriate outputs. In all of these cases, the functions of the truth table, state machine, or Boolean equations could be modified by simply exchanging the programmable device with an upgraded version.

a b c	w x y
0 0 0	0 1 1
0 0 1	1 0 0
0 1 0	0 0 0
0 1 1	1 1 0
1 0 0	1 0 1
1 0 1	0 0 1
1 1 0	1 0 0
1 1 1	1 0 1

Old PROM

New PROM

a b c	w x y
0 0 0	1 0 1
0 0 1	1 1 0
0 1 0	0 1 0
0 1 1	1 0 1
1 0 0	0 1 0
1 0 1	0 0 1
1 1 0	1 1 1
1 1 1	1 0 0

**Figure 21.2: Reconfigurable hardware:
PROM as a hardware truth table**

Another technique would be to use non-volatile memory devices to store firmware programs for use by a microprocessor or a microcontroller. An example could be a set of instructions used by a microprocessor to play a tune such as the National Anthem on a musical door chime. Different versions of the PROM could be used to allow the product to be marketed in different countries (hopefully with the correct anthem). Additionally, in the case of those countries that count revolution as a national sport, the product could be easily reconfigured to reflect the "tune of the day."

The non-volatile memory examples introduced above could use PROM, EPROM, E^2PROM, or FLASH devices; similarly, the programmable logic device examples could employ PLD, EPLD, E^2PLD, or FLASH-PLD components. In all of these cases, from a board-level perspective, the board itself would be classified as reconfigurable. However, from a device-level viewpoint, PROMs would fall into the category of *configurable*, while their more sophisticated cousins, EPROM, E^2PROM, and FLASH would be categorized as *reconfigurable* (similarly in the case of PLDs versus EPLDs, E^2PLDs, and FLASH-PLDs). Additionally, E^2-based and FLASH-based components may be referred to as *in-system programmable (ISP)*, because they can be reprogrammed while remaining resident on the circuit board.

Dynamically Reconfigurable Hardware

The advent of SRAM-based FPGAs in the late 1980s and early 1990s presented a new capability to the electronics fraternity: dynamically reconfigurable hardware, which means designs that can be reconfigured "on-the-fly."

FPGAs contain a large number of diverse logic gates and registers, which can be connected together in widely different ways to achieve a desired function. SRAM-based variants augment the capabilities of standard FPGAs by allowing new configuration data to be down-loaded into the device by the main system in a fraction of a second. In the case of these devices, a few of the external pins are dedicated to the task of loading the data, including an *enable* control and *clock* and *data* inputs. When the *enable* control is placed in its active state, edges on the *clock* are used to load the device's SRAM with a stream of 0s and 1s which are presented to the serial *data* input. Although all of the logic gates and SRAM cells are created on the surface of a single piece of silicon substrate, it may be useful to visualize the device as comprising two distinct strata: the logic gates and the programmable SRAM "switches."

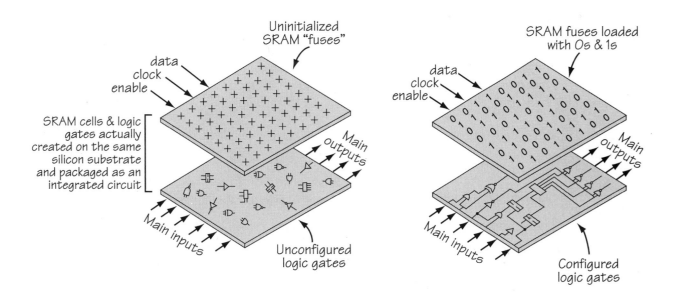

Figure 21.3: Dynamically reconfigurable hardware: SRAM-based FPGAs

The versatility of these devices opens the floodgates to a wealth of possibilities. For example, the creation of circuit boards whose interfaces can be configured to meet a variety of communications protocols, or devices acting as *digital signal processors (DSPs)*, whose core algorithms can be modified to process data in a variety of ways. As alternative protocols become available or improved algorithms are invented, the patterns that are used to configure the FPGAs can be modified to take full advantage of these new developments.

However, the true power of these devices, which are referred to as *in-circuit reconfigurable (ICR)*, resides in their ability to be reconfigured on-the-fly. For example, when a system is first turned on, it might configure all of the FPGAs to perform diagnostic functions, both on themselves and on the circuit board. After the diagnostic checks have been completed, the system can dynamically reconfigure the FPGAs to fulfill the main function of the design.

Another example is illustrated by the Tomahawk cruise missile, which uses one technique to control itself while flying over water and another while soaring over land. When the Tomahawk crosses the boundary from water to land or vice versa, it causes its FPGAs to be dynamically reconfigured, changing from water-navigation mode to land-navigation mode in a fraction of a second.[7]

Dynamically Reconfigurable Interconnect

Wonderful as all of the above is, even these techniques only scratch the surface of the possibilities offered by today's emerging technologies. Designers would ideally like to create board-level products that can be reconfigured to perform radically improved, or completely different functions from the ones that they were originally designed for. The solution is to be able to dynamically configure the board-level connections between devices.

---

[7] Of course, some of us might take the view that it is inherently unwise to have an armed cruise missile "bombing" around the sky in a mindless state while it reprograms its own brain!

A new breed of devices offer just this capability: *field-programmable interconnect devices (FPIDs)*, which may also be known as *field-programmable interconnect chips (FPICs)*.[8] These devices, which are used to connect logic devices together, can be dynamically reconfigured in the same way as standard SRAM-based FPGAs. Because each FPID may have around 1,000 pins, only a few such devices are typically required on a circuit board:

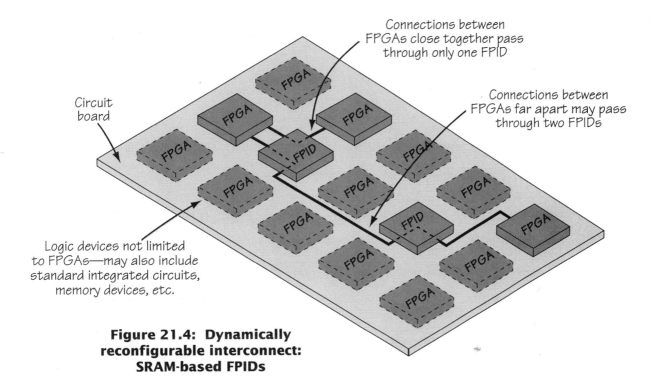

Connections between FPGAs close together pass through only one FPID

Connections between FPGAs far apart may pass through two FPIDs

Circuit board

Logic devices not limited to FPGAs—may also include standard integrated circuits, memory devices, etc.

Figure 21.4: Dynamically reconfigurable interconnect: SRAM-based FPIDs

In fact, the concepts discussed here are not limited to board-level implementations. Any of the technologies discussed thus far may also be implemented in hybrids and multichip modules. Additionally, within the foreseeable future, ultra-large-scale ASIC devices will become available, which combine microprocessor cores, blocks of memory, and communication functions with embedded FPGA-style and FPID-style functions.

---

[8] FPIC is a trademark of Aptix Corporation.

Virtual Hardware

The main limitation with the majority of SRAM-based FPGAs is that it is necessary to load the entire device. Apart from anything else, it is usually necessary to halt the operation of the entire circuit board while these devices are being reconfigured. Additionally, the contents of any registers in the FPGAs are irretrievably lost during the process.

To address these issues, a new generation of FPGAs were introduced around the beginning of 1994. In addition to supporting the dynamic reconfiguration of *selected portions* of the internal logic, these devices also feature:

a) No disruption to the device's inputs and outputs.

b) No disruption to the system-level clocking.

c) The continued operation of any portions of the device that are not undergoing reconfiguration.

d) No disruption to the contents of internal registers during reconfiguration, even in the area being reconfigured.

The latter point is of particular interest, because it allows one instantiation of a function to hand over data to the next function. For example, a group of registers may initially be configured to act as a binary counter. Then, at some time determined by the main system, the same registers may be reconfigured to operate as a linear feedback shift register (LFSR),[9] whose seed value is determined by the final contents of the counter before it was reconfigured.

Although these devices are evolutionary in terms of technology, they are *revolutionary* in terms of the potential they offer. To reflect their new capabilities, appellations such as *virtual hardware, adaptive hardware,* and *Cache Logic*[10] are beginning to emerge. Because it appears likely that this nomenclature may quickly work its way into mainstream usage, it is appropriate to take a few moments to explain the roots of these terms.

---

[9] Linear feedback shift registers (LFSRs) are introduced in detail in Appendix F.

[10] Cache Logic is a trademark of Atmel Corporation, San Jose, CA, USA, to whom thanks are due for the information that they provided at such short notice.

The phrase *"virtual hardware"* is derived from its software equivalent, *"virtual memory,"* and both are used to imply something that is not really there. In the case of virtual memory, the computer's operating system pretends that it has access to more memory than is actually available. For example, a program running on the computer may require ten Mbytes to store its data, but the computer may have only five Mbytes of memory available. To get around this problem, whenever the program attempts to access a memory location that does not physically exist, the operating system performs a sleight-of-hand and exchanges some of the contents in the memory with data on the hard disk. Although this practice, known as *swapping,* tends to slow things down, it does allow the program to perform its task, without having to wait while someone runs down to the store to buy some more memory chips.

Similarly, the phrase *"Cache Logic"* is derived from its similarity to the concept of *"Cache Memory,"* in which high-speed, expensive SRAM is used to store active data,[11] while the bulk of the data resides in slower, lower-cost memory devices such as DRAM.

In fact, the concepts behind virtual hardware are actually quite easy to understand. Each large macro-function in a device is usually formed by the combination of a number of smaller micro-functions such as counters, shift registers, and multiplexers. Two things become apparent when a group of macro-functions are divided into their respective micro-functions. First, functionality overlaps, and an element such as a counter may be used several times in different places. Second, there is a substantial amount of *functional latency,* which means that, at any given time, only a portion of the micro-functions are active. To put this another way, relatively few micro-functions are in use during any given clock cycle. Thus, the ability to dynamically reconfigure individual portions of a virtual hardware device means that a relatively small amount of logic can be used to implement a number of different macro-functions.

---

[11] In this context, *"active data"* refers to data or instructions that a program is currently using, or which the operating system believes that the program will want to use in the immediate future.

By tracking the occurrence and usage of each micro-function, then consolidating functionality and eliminating redundancy, virtual hardware devices can perform far more complex tasks than they would appear to have logic gates available. For example, in a complex function requiring 10,000 equivalent gates, only 2,000 gates may be active at any one time. Thus, by storing, or caching, the functions implemented by the extra 8,000 gates, a small, inexpensive 2,000-gate device can be used to replace a larger, more expensive 10,000-gate component:

Figure 21.5: Virtual hardware

In fact, it is even possible to "compile" new design variations in real-time, which may be thought of as dynamically creating subroutines in hardware! Hence the phrase "*adaptive hardware*" referred to above.

Three-Dimensional Molded Interconnect

For many years designers have wished for the ability to create robust three-dimensional (3-D) circuit boards for use with products such as hand-held cellular telephones, radios, and calculators. In addition to providing the interconnect, these 3-D circuit boards would also act as the product's package (from another point of view, designers want to be able to create a 3-D product package that also acts as the interconnect). In addition to reducing the product's size and weight, there can also be significant benefits in terms of cost and manufacturability.

A number of processes to create 3-D circuit boards have emerged over the years, but none have gained any significant level of commercial acceptance. One hopeful contender involved creating a standard two-dimensional (2-D) circuit board and then molding it into a 3-D shape. This process achieved some acceptance for simple product packages, such as those for power supplies, with limited aesthetic requirements. However, the process is not suitable for products that require an aesthetic appeal and ergonomic shapes involving complex surfaces.

Another technique involved the prefabrication of tracks on the inside surface of a mold. The process of injecting plastic into the mold caused the tracks to become an integral part of the packaging. But this process had its own problems—not the least, creating the interconnects on the 3-D surface of the mold in the first place.

However, a new development in 3-D photo-imaging technology in the early 1990s has reawakened interest in the injection molding technique.[12] The process commences with the injection molding of a plastic material capable of withstanding high enough temperatures to undergo reflow-soldering or vapor-phase soldering processes. In addition to the physical shape of the product, it is also possible to mold in features such as holes, ribs, recesses, standoffs, and chamfered edges. These features offer substantial savings compared to their equivalent drilling, routing, and grinding operations, and also reduce part counts and labor.

---

[12] Thanks to Faldering Design Services, Hamden, CT, USA, for the information on this process.

A 3-D photo-tool, or mask, is formed by molding a PVC-based material to conform exactly to the contours of the plastic part. Next, a computer-controlled laser is used to draw an image of the desired circuit onto the photo-tool.

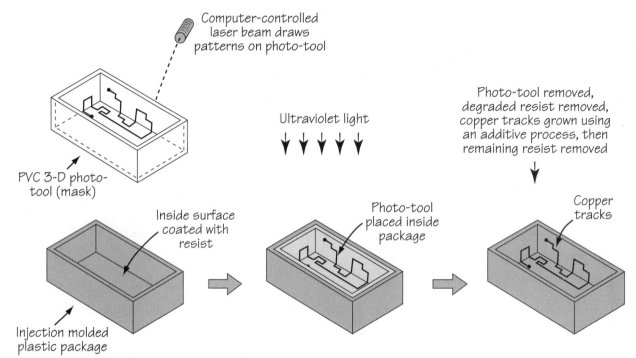

Figure 21.6: Three-dimensional molded interconnect

The remainder of the process is very similar to that for a standard circuit board, and can therefore leverage off existing technology. The surface of the plastic is covered by a special resist, the photo-tool is inserted, the whole assembly is exposed to ultraviolet light, the degraded resist is removed, and copper tracks are grown using an additive process.

One of the major problems with the process at this time is the lack of appropriate computer-aided design tools. Existing tools are geared to standard 2-D circuit boards and do not have the ability to transform 2-D designs into a 3-D space. Sadly, there is not yet a large enough market potential to cause this position to change. However, this technology is still in its infant stage and, with sufficient market interest, the requisite electromechanical toolsets will become available.

Optical Interconnect

Electronics systems exhibit ever-increasing requirements to process ever-increasing quantities of data at ever-increasing speeds. Interconnection technologies based on conducting wires are fast becoming the bottleneck that limits the performance of electronic systems.

To relieve this communications bottleneck, a wide variety of opto-electronic interconnection techniques are undergoing evaluation. In addition to the extremely fast propagation of data,[13] optical interconnects offer greater signal isolation, reduced sensitivity to electromagnetic interference, and a far higher bandwidth than do conducting wires.

Fiber-Optic Interconnect

The fibers used in fiber-optic systems are constructed from two different forms of glass (or other materials) with different refractive indices. These fibers, which are finer than a human hair, can be bent into weird and wonderful shapes without breaking. When light is injected into one end of the fiber, it repeatedly bounces off the interface between the two glasses, undergoing almost total internal reflection with minimal loss, until it reemerges at the other end:

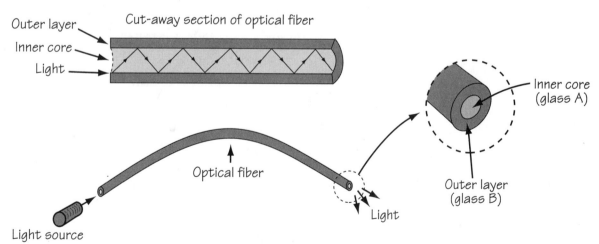

Figure 21.7: Light propagating through an optical fiber

---

[13] Light travels at 299,792,458 meters per second in a vacuum; thus, a beam of light would take only approximately 2.6 seconds to make a round trip from the earth to the moon and back again!

Experimental systems using fiber-optic interconnect have been evaluated at all levels of a system—for example, to link bare die in a multichip module:

Figure 21.8: Fiber-optic interconnect: linking bare die in a multichip module

The transmitting device employs a surface-emitting laser-diode, which is constructed along with the transistors and other components on the integrated circuit's substrate. The receiving device uses a photo-transistor to convert the incoming light back into an electrical signal.

Each die can support numerous transmitters and receivers, which can be located anywhere on the surface of their substrates. However, there are several problems with this implementation, including the difficulty of attaching multiple fibers, the difficulties associated with repair and rework (replacing a defective die), and the physical space occupied by the fibers. Although the individual fibers are extremely thin, multichip modules may require many thousands of connections. Additionally, in this form, each optical fiber can only be used to connect an individual transmitter to an individual receiver.

At the board level, a variation of discrete wired technology[14] has been developed, in which optical fibers are ultrasonically bonded into the surface of a board. When combined with chip-on-board techniques, this process

---

[14] Discrete wired technology was introduced in Chapter 18.

offers significant potential for the future. In fact, due to the way in which the optical fibers are connected to the die, it is possible for an individual transmitter to be connected to multiple receivers. Unfortunately, the techniques used to achieve this may not be disclosed here, because the author has been sworn to secrecy using arcane oaths and strange handshakes, and an Englishman's word is his bond!

Last, but not least, optical fibers may be used to provide intra-board connections, which may be referred to as *optical backplanes*. Once again, discrete wire technology, modified to accommodate optical fibers, may be employed to create backplanes. Alternatively, the boards may be mounted in a rack without an actual backplane, and groups of optical fibers may be connected into special couplers.

Figure 21.9: Fiber-optic inter-connect: optical backplanes

Each optical fiber from a transmitter is connected into a coupler, which amplifies the optical signal and can re-transmit it to multiple receivers. This form of backplane offers great latitude in regard to the proximity of the boards. In fact, boards connected in this way can be separated by as much as tens of meters.

Free-Space Interconnect

With the free-space technique, a laser-diode transmitter communicates directly with a photo-transistor receiver without employing an optical fiber. Consider a free-space technique used to link bare die mounted on the substrate of a multichip module:

**Figure 21.10:
Free-space interconnect: linking
bare die in a multichip module**

In this case, the transmitters are constructed as side-emitting laser diodes along the upper edges of the die; similarly with the photo-transistors on the receivers. Each die may contain multiple transmitters and receivers. The free-space technique removes some of the problems associated with its fiber-optic equivalent: there are no fibers to attach and replacing a defective die is easier.

However, as for optical fibers, each transmitter can still only be used to connect to an individual receiver. Additionally, the free-space technique has its own unique problems: the alignment of the devices is critical and there are *thermal tracking* issues. When a laser-diode is turned on, it rapidly cycles from ambient temperature to several hundred degrees Celsius. The heat generated by an individual laser-diode does not greatly affect the die, because each diode is so small. But the cumulative effect of hundreds of such diodes does affect the die, causing it to expand and therefore disturbing the alignment of the transmitter-receiver pairs.

A somewhat more significant consideration is that, unless the system is constructed as an array of identical die, each die has to be fabricated with unique arrangements of transmitters to interface to its surrounding die. This is not unlike an extremely complex jigsaw, and is probably the key deficiency that will prevent this technique from maturing into a commercially viable proposition.

Guided-Wave Interconnect

Another form of optical interconnect that is receiving significant interest is that of guided-wave, whereby optical waveguides are fabricated directly on the substrate of a multichip module. These waveguides can be created using variations on standard opto-lithographic thin-film processes. One such process involves the creation of silica waveguides:[15]

Figure 21.11: Guided-wave interconnect: silica waveguides

---

[15] Thanks for the information on silica waveguides go to Dr. Terry Young of the GEC-Marconi Research Center, Chelmsford, Essex, England, with apologies for all the extremely technical details that were omitted here.

Using a flipped-chip mounting technique, the surface-emitting laser-diodes and photo-transistors on the component-side of the die are pointed down towards the substrate. One major advantage of the waveguide technique is that the waveguides can be constructed with splitters, thereby allowing a number of transmitters to drive a number of receivers. On the down-side, it is very difficult to route one waveguide over another, because the crossover point tends to act like a splitter and allows light from the waveguides to "leak" into each other.

An alternative form of waveguide technology, *photo-imagable polyimide interconnect*, which was announced towards the tail-end of 1993, is of particular interest in the case of multichip modules of type MCM-D. As you may recall from Chapter 20, MCM-D devices are "*. . . ceramic, glass, or metal substrates that are covered with a layer of dielectric material such as polyimide. The dielectric coat is used to modify the substrate's capacitive characteristics and tracks are created on the surface of the dielectric using thin-film processes.*"

When exposed to light passed through an appropriate mask, a layer of photo-imagable polyimide can be imprinted with patterns in a similar way to exposing a photograph. After being developed, the polyimide contains low-loss optical waveguides bounded by relatively opaque reflective surfaces (Figure 21.12).

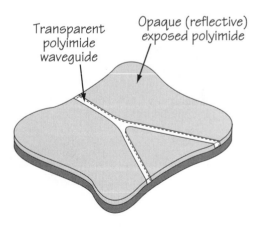

Figure 21.12: Guided-wave interconnect: photo-imagable polyimide waveguides

Apart from its inherent simplicity, one of the beauties of this technique is that both the exposed and unexposed areas of polyimide have almost identical dielectric constants. Thus, in addition to leveraging off existing technology, the polyimide waveguides have relatively little impact on any thin-film metallization tracking layers that may be laid over them.

This technique is currently finding its major audience in designers of multichip modules, but it is also being investigated as a technique for multilayer circuit boards. Future circuit boards could therefore be fabricated as a mixture of traditional copper interconnect and very high speed optical interconnect.

Holographic Interconnect

Unfortunately, all of the optical interconnection technologies introduced above have their own unique problems and limitations. The technique that shows most potential is that of guided-waves, but even this has the problem of routing waveguides over each other. Additionally, both waveguides and optical fibers share a common shortcoming, the fact that the light tends to bounce around an awful lot. In fact, the light bounces around so much that it actually travels approximately four to six times the straight-line distance between the transmitter and the receiver.

Quite apart from these concerns, it is doubtful if any of the above techniques will be capable of dealing with the sheer number of interconnection paths that will be required. A new contender that is rising to the challenge is holographic interconnect. Using the term *"holographic"* in this context may at first seem a little strange, because *holography* is traditionally considered to be a method of obtaining three-dimensional images known as *holograms* (from the Greek: *holos* meaning "whole" and *gram* meaning "message"). However, the term holographic is appropriate, because this form of interconnect is actually based on a three-dimensional image.

In the case of holographic interconnect for a multichip module, the process commences with an extremely thin slice of quartz, into which sophisticated patterns are cut using a laser. The quartz slice is then mounted approximately 1 mm above the die, and a face-surface mirror[16] is mounted approximately 2 cm above the quartz:

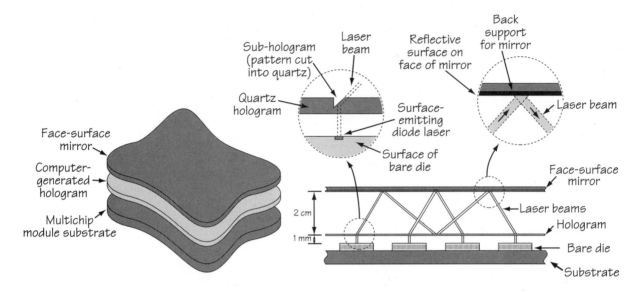

Figure 21.13: Holographic interconnect:
multichip module

When a surface-emitting laser-diode on one of the die is turned on, it transmits a laser beam straight up into one of the patterns in the quartz. This pattern causes the laser beam to be deflected so that it bounces backwards and forwards between the face-surface mirror and the quartz. When the laser hits the point on the quartz directly above a photo-transistor on the receiving die, another pattern cut into the quartz causes it to be deflected back through the quartz and down into the receiver.

Note that the patterns cut into the quartz in the illustration above are gross simplifications. Additionally, there are typically two patterns cut into the quartz over the receiver, one on each side of the slice. The pattern on the upper side of the slice captures the laser beam and deflects it into the slice, and the pattern on the lower side then captures it again and redirects

---

[16] The term "face-surface" refers to the fact that the reflective coating is located on the face of the mirror, and not behind it as would be the case in a mirror you look into when you shave or apply your makeup (or both, depending on how liberated you are).

it out of the slice down onto the receiver. These patterns also perform the same function as a lens, focusing the beam precisely onto the receiver.

The angles of the patterns cut into the quartz are precisely calculated, as are the number of reflections between the face-surface mirror and the quartz, thereby ensuring that each laser beam exactly hits the target for which it is intended and no other. All of these calculations are performed by a computer, and the laser used to cut the three-dimensional patterns into the quartz is also controlled by a computer. Thus, the slice of quartz is referred to a *computer-generated hologram (CGH)*, and the individual patterns above each transmitter and receiver are known as sub-holograms.

In fact, the laser beam actually undergoes relatively few reflections between the face-surface mirror and the quartz; far fewer than an equivalent beam transmitted through a waveguide or an optical fiber. Thus, the time taken for a signal to propagate from a transmitter to a receiver, known as the *time-of-flight*, is actually close to that for free-space interconnect. But the clever part is still to come. The sub-hologram above a laser-diode can be created in such a way that it splits the laser beam into sub-beams, each of which can target a different receiver:

**Figure 21.14: Holographic interconnect:
splitting a laser beam**

For the sake of clarity, the previous illustration shows a hexagonal sub-hologram with equal angles between the vertices, and equal angles of incidence for each face. However, the angles between the vertices can vary, and each face can have a different angle of incidence. Furthermore, the sub-holograms above receivers can be correspondingly complex, and capable of receiving signals from multiple transmitters.

Finally, in addition to multichip module applications, holographic interconnect is also undergoing evaluation as a possible backplane technology for the interconnection of circuit boards. These backplanes, known as *holobackplanes*, offer an additional, amazingly powerful capability. It is technically feasible to create multiple holograms in the backplane's equivalent of the quartz slice. Thus, by moving the holographic slice a fraction of a millimeter in a direction parallel to the face-surface mirror, the interconnection pattern could be reconfigured and a completely new set of board-to-board interconnects could be established. The mind boggles!

Optical Memories

It has been estimated that the total sum of human knowledge is doubling approximately every ten years. Coupled with this, the amount of information that is being generated, stored, and accessed is increasing at an exponential rate. This is driving the demand for fast, cheap memories that can store gigabits, or even terabits, of data.

One medium with the potential to cope with this level of data density is optical storage. Among many other techniques, evaluations are being performed on extremely thin layers of glass-based materials,[17] which are doped with organic dyes or rare-earth elements. Using a technique known as *photochemical hole-burning (PHB)*, a laser in the visible waveband is directed at a microscopic point on the surface of the glass (Figure 21.15).

If the laser is weak, its light will pass through the glass without affecting it and reappear at the other side. If the laser is stronger (but not intense enough to physically damage the glass), electrons in the glass will be excited by the light. The electrons can be excited such that they change the absorp-

---

[17] One such material, boric-acid glass, is also widely used in heat-resistant kitchen ware!

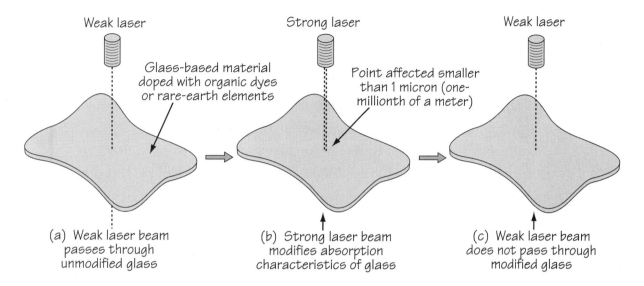

Figure 21.15: Optical memories: photochemical hole-burning (PHB)

tion characteristics of that area of the glass and leave a band, or hole, in the absorption spectrum. To put this another way, if the weak laser beam is redirected at the same point on the glass surface, its light would now be absorbed and would *not* reappear at the other side of the glass.

Thus, depending on whether or not the light from the weaker beam passes through the glass, it can be determined whether or not that point has been exposed to the strong laser. This means that each point can be used to represent a binary 0 or 1. Because the point affected by the laser is so small, this process can be replicated millions upon millions of times across the surface of the glass.

If a point occurs at one micron (one millionth of a meter) intervals, then it is possible to store 100 megabits per square centimeter, but this still does not come close to the terabit storage that will be required. However, it turns out that each point can be "multiplexed" and used to store many bits of information. A small change in the wavelength of the laser can be used to create another hole in a different part of the spectrum. In fact, 100× multiplexing has been achieved, where each point on the glass was used to store 100 bits of data at different wavelengths. Using 100× multiplexing offers a data density of 10 gigabits per square centimeter, and even higher levels of multiplexing may be achieved in the future!

Protein Switches and Memories

Another area receiving a lot of interest is that of switches and memories based on proteins.[18] Organic molecules have a number of useful properties, not the least that their structures are intrinsically *"self healing"* and reject contamination. Also, in addition to being extremely small, many organic molecules have excellent electronic properties. Unlike metallic conductors, they transfer energy by moving electron excitations from place to place rather than relocating entire electrons. This can result in switching speeds that are orders of magnitude faster than their semi-conductor equivalents.

Some proteins react to electric fields, while others respond to light. For example, there is a lot of current interest in the protein Rhodopsin, which is used by certain photosynthetic bacteria to convert light into energy. The bacteria that contain Rhodopsin are the ones that cause ponds to turn red, and their saltwater cousins are responsible for the purple tint which is sometimes seen in San Francisco Bay.

Laser beam at frequency "A"

Proteins change color at intersection point of two laser beams

Laser beam at frequency "B"

3-D cube of optically-responsive proteins suspended in transparent polymer

Figure 21.16: Protein memories: cubic arrays of light-sensitive proteins

---

[18] A complex organic molecule formed from chains of amino acids, which are themselves formed from combinations of certain atoms, namely: carbon, hydrogen, nitrogen, oxygen, usually sulfur, and occasionally phosphorus or iron. Additionally, the chain of amino acids *"folds in on itself"* forming an extremely complex 3-D shape.

In certain cases, light from a laser can be used to cause such optically responsive proteins to switch from one state to another (which they do by changing color) and back again. Additionally, some varieties of proteins are only responsive to the influence of two discrete frequencies. This feature is extremely attractive, because it offers the possibility of three-dimensional optical protein memories.

Experiments have been performed, in which 3-D cubes have been formed as ordered arrays of such bi-frequency proteins suspended in transparent polymers. If the protein were affected by a single laser, then firing a beam into the cube would result in a line of proteins changing state. But in the case of bi-frequency proteins, two lasers mounted at 90° to each other can be used to address individual points in the 3-D space (Figure 21.16).

By only slightly enhancing the technology available today, it may be possible to store as much as 20 gigabits in a 1 cm × 1 cm × 1 cm cube of such material, where even one gigabit would be equivalent to 1,250 of today's 16 megabit RAM devices!

Electromagnetic Transistor Fabrication

For some time it has been known that the application of strong electromagnetic fields to special compound semiconductors can create structures that behave like transistors. The original technique was to coat the surface of a semiconductor substrate with a layer of dopant material, and then to bring an extremely strong, concentrated electromagnetic field in close proximity.

The theory behind this original technique was that the intense field caused the electromigration of the dopant into the substrate. However, much to everyone's surprise, it was later found that this process remained effective even without the presence of the dopant!

Strange as it may seem, nobody actually understands the mechanism that causes this phenomenon. Physicists currently suspect that the strong electromagnetic fields cause microscopic native defects in the crystals to migrate through the crystal lattice and cluster together.

Heterojunction Transistors

If there is one truism in electronics, it is that *"faster is better,"* and a large proportion of research and development funds are invested in increasing the speed of electronic devices.

Ultimately there are only two ways to increase the speed of semiconductor devices. The first is to reduce the size of the structures on the semiconductor, thereby obtaining smaller transistors that are closer together. The second is to use alternative materials that inherently switch faster. However, although there are a variety of semiconductors, such as gallium arsenide, that offer advantages over silicon for one reason or another, silicon is cheap, readily available, and relatively easy to work with. Additionally, the electronics industry has billions of dollars invested in silicon-based processes.

For these reasons, speed improvements have traditionally been achieved by making transistors smaller. However, it is becoming apparent that we are reaching the end of this route using conventional technologies. At one time, the limiting factors appeared to be simple process limitations: the quality of the resist, the ability to manufacture accurate masks, and the features that could be achieved with the wavelength of ultraviolet light. Around 1990, when structures with dimensions of 1.0 microns first became available, it was believed that structures of 0.5 microns would be the effective limit that could be achieved with opto-lithographic processes, and that the next stage would be a move to X-ray lithography. However, there have been constant improvements in the techniques associated with mask fabrication, optical systems and lenses, servo motors and positioning systems, and advances in chemical engineering such as chemically-amplified resists.[19] The combination of all these factors means that it is now considered feasible to achieve structures as small as 0.1 microns by continuing to refine existing processes.

---

[19] In the case of a chemically-amplified resist, the application of a relatively small quantity of ultraviolet light stimulates the formation of chemicals in the resist which accelerates the degrading process. This reduces the amount of ultraviolet light which is required to degrade the resist and allows the creation of finer features with improved accuracy.

However, there are other considerations. The speed of a transistor is strongly related to its size, which affects the distance electrons have to travel. Thus, to enable transistors to switch faster, technologists have concentrated on reducing size, a strategy which is commonly referred to as *scaling*. However, while scaling reduces the size of structures, it is necessary to maintain certain levels of dopants to achieve the desired effect. This means that, as the size of the structures is reduced, it is necessary to increase the concentration of dopant atoms. Increasing the concentration beyond a certain level causes leakage, resulting in the transistor being permanently ON and therefore useless. Thus, technologists are increasingly considering alternative materials and structures.

An interface between two regions of semiconductor having the same basic composition but opposing types of doping is called a *homojunction*. By comparison, the interface between two regions of dissimilar semiconductor materials is called a *heterojunction*. Homojunctions dominate current processes because they are easier to fabricate. However, the interface of a heterojunction has naturally occurring electric fields which can be used to accelerate electrons, and transistors created using heterojunctions can switch much faster than their homojunction counterparts of the same size.

One form of heterojunction that is attracting a lot of interest is found at the interface between silicon and germanium. Silicon and germanium are in the same family of elements and have similar crystalline structures which, in theory, should make it easy to combine them but, in practice, is a little more difficult. A process currently being evaluated is to create a standard silicon wafer with doped regions, and then to grow extremely thin layers of a silicon-germanium alloy where required.

The two most popular methods of depositing these layers are *chemical vapor deposition (CVD)* and *molecular beam epitaxy (MBE)*. In the case of chemical vapor deposition, a gas containing the required molecules is converted into a *plasma*[20] by heating it to extremely high temperatures using microwaves. The plasma carries atoms to the surface of the wafer where they

---

[20] A gaseous state in which the atoms or molecules are dissociated to form ions.

are attracted to the crystalline structure of the substrate. This underlying structure acts as a template. The new atoms continue to develop the structure to build up a layer on the substrate's surface:

Figure 21.17: Heterojunction transistors: chemical vapor deposition (CVD)

By comparison, in the case of molecular beam epitaxy, the wafer is placed in a high vacuum, and a guided beam of ionized molecules is fired at it, effectively allowing molecular-thin layers to be "painted" onto the substrate where required.[21]

Ideally, such a heterojunction would be formed between a pure silicon substrate and a pure layer of germanium. Unfortunately, germanium atoms are approximately 4% larger than silicon atoms, the resulting crystal lattice cannot tolerate the strains that develop, and the result is defects in the structure. In fact, millions of minute inclusions occur in every square millimeter, preventing the chip from working. Hence, the solution of growing a layer of silicon-germanium alloy, which relieves the stresses in the crystalline structure, thereby preventing the formation of inclusions (Figure 21.18).

---

[21] Molecular beam epitaxy is similar to *electron beam epitaxy (EBE)*, in which the wafer is first coated with a layer of dopant material before being placed in a high vacuum. A guided beam of electrons is fired at the wafer causing the dopant to be driven into it.

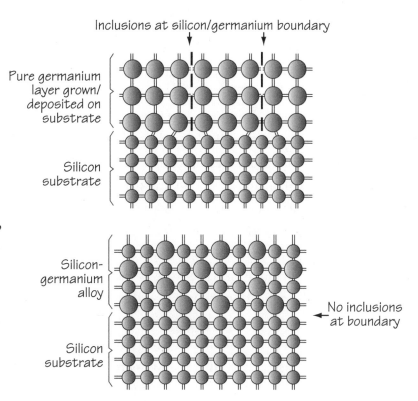

=○= Silicon atom

=◉= Germanium atom

Heterojunctions offer the potential to create transistors that switch as fast, or faster, than those on gallium arsenide, but use significantly less power. Additionally, they have the advantage of being able to be produced on existing fabrication lines, thereby preserving the investment and leveraging current expertise in silicon-based manufacturing processes.

Figure 21.18: Heterojunction transistors: silicon-germanium alloy

Diamond Substrates

As was noted in the previous section, there is a constant drive towards smaller, more densely packed transistors switching at higher speeds. Unfortunately, packing the little devils closer together and cracking the whip to make them work faster substantially increases the amount of heat that they generate. Similarly, the increasing utilization of optical interconnect relies on the use of laser diodes, but today's most efficient laser diodes only convert 30% to 40% of the incoming electrical power into an optical output, while the rest emerges in the form of heat. Although each laser diode is relatively small (perhaps as small as only 500 atoms in diameter), their heating effect becomes highly significant when tens of thousands of them are performing their version of Star Wars.

And so we come to diamond, which derives its name from the Greek *adamas*, meaning "invincible." Diamond is famous as the hardest substance known, but it also has a number of other interesting characteristics: it is a better conductor of heat at room temperatures than any other material,[22] in its pure form it is a good electrical insulator, it is one of the most transparent materials available, and it is extremely strong and noncorrosive. For all of these reasons, diamond would form an ideal substrate material for multichip modules.[23]

In addition to multichip modules, diamond has potential for a variety of other electronics applications. Because diamond is in the same family of elements as silicon and germanium, it can function as a semiconductor and could be used as a substrate for integrated circuits. In fact, in many ways, diamond would be far superior to silicon: it is stronger, it is capable of withstanding high temperatures, and it is relatively immune to the effects of radiation (the bane of components intended for nuclear and space applications). Additionally, due to diamond's high thermal conductivity, each die would act as its own heat sink and would rapidly conduct heat away. It is believed that diamond-based devices could switch up to 50 times faster than silicon and operate at temperatures over 500°C.

Unfortunately, natural diamond is extremely expensive and, if you should happen to find one of the little beauties, the last thing that would come to mind would be to chop it up into thin slices for electronics applications! However, there are a number of methods for depositing or growing diamond crystals, one of the most successful being chemical vapor deposition (CVD), which was introduced in the previous discussions. In this CVD process, microwaves are used to heat mixtures of hydrogen and hydrocarbons into a plasma, out of which diamond films nucleate and form on suitable substrates. Although the plasma chemistry underlying this phenomena is not fully understood, polycrystalline diamond films can be nucleated on a wide variety of materials, including metals such as titanium, molybdenum, and tungsten, ceramics, and other hard materials such as quartz, silicon, and sapphire.

---

[22] Diamond can conduct five times as much heat as copper, which is the second most thermally-conductive material known.

[23] Actually, other exotic substrates are also of interest to electronic engineers, including sapphire, which is of particular use in microwave applications.

CVD processes work by growing layers of diamond directly onto a substrate. A similar, more recent, technique, known as *chemical vapor infiltration (CVI)*,[24] commences by placing diamond powder in a mold. Additionally, thin posts, or columns, can be preformed in the mold, and the diamond powder can be deposited around them. When exposed to the same plasma as used in the CVD technique, the diamond powder coalesces into a polycrystalline mass. After the CVI process has been performed, the posts can be dissolved leaving holes through the diamond for use in creating vias. CVI processes can produce diamond layers twice the thickness of those obtained using CVD techniques at a fraction of the cost.

An alternative, relatively new technique for creating diamond films involves heating carbon with laser beams in a vacuum. Focusing the lasers on a very small area generates extremely high temperatures, which rip atoms away from the carbon and also strip away some of their electrons. The resulting ions fly off and stick to a substrate placed in close proximity. Because the lasers are tightly focused, the high temperatures they generate are localized on the carbon, permitting the substrate to remain close to room temperature. Thus, this process can be used to create diamond films on almost any substrate, including semiconductors, metals, and plastics.

The number of electrons stripped from the carbon atoms varies, allowing their ions to reform in *nanophase* diamond structures which have never been seen before. Nanophase materials are a new form of matter which was only recently discovered, in which small clusters of atoms form the building blocks of a larger structure. These structures differ from those of naturally occurring crystals, in which individual atoms arrange themselves into a lattice. In fact, it is believed that it may be possible to create more than thirty previously unknown forms of diamond using these techniques.

Last, but not least, in the late 1980s, a maverick inventor called Ernest Nagy[25] invented a simple, cheap, and elegant technique for creating thin diamond films. Nagy's process involves treating a soft pad with diamond

---

[24] Thanks go to Crystallume, Menlo Park, CA, USA, for the information on their CVD and CVI processes.

[25] Nagy, whose full name is Ernest Nagy de Nagybaczon, was born in 1942 in Hungary. He left as a refugee in the 1956 uprising and now lives in England.

powder, spinning the pad at approximately 30,000 revolutions per minute, and maintaining the pad in close contact with a substrate. Although the physics underlying the process is not fully understood, diamond is transferred from the pad to form a smooth and continuous film on the substrate. The diamond appears to undergo some kind of phase transformation, changing from a cubic arrangement into a hexagonal form with an unusual structure. Interestingly enough, Nagy's technique appears to work with almost any material on almost any substrate!

All of the techniques described above result in films that come respectfully close, if not equal, to the properties of natural diamond in such terms as heat conduction. Thus, these films are highly attractive for use as substrates in multichip modules. However, the unusual diamond structures that are created fall short of the perfection required for them to be used as a substrate suitable for the fabrication of transistors.

Substrates for integrated circuits require the single, large crystalline structures found only in natural diamond. Unfortunately, there are currently no known materials onto which a single-crystal diamond layer will grow, with the exception of single crystal diamond itself (which sort of defeats the point of doing it in the first place). The only answer appears to be to modify the surface of the substrate onto which the diamond layer is grown, and many observers believe that this technology may be developed in the near future. If it does prove possible to create consistent, single-crystal diamond films then, in addition to being *"a girl's best friend,"* diamonds would also become *"an engineer's biggest buddy."*

Chip-On-Chip (COC)

The intra-chip connections linking bare die on a multichip module are a source of fairly significant delays. One obvious solution is to mount the die as closely together as possible, thereby reducing the lengths of the tracks and the delays associated them. However, each die can only have eight others mounted in close proximity on a 2-D substrate. The solution is to proceed into three dimensions. Each die is very thin and, if they are mounted on top of each other, it is possible to have over a hundred die forming a 3-D cube (Figure 21.19).

Figure 21.19: Chip-on-chip: die organized as 3-D cubes

One problem with this chip-on-chip (COC) technique is the amount of heat that is generated, which drastically affects the inner layers forming the cube. This problem could be eased by constructing the die out of diamond as discussed above. Another problem with traditional techniques is that any tracks linking the die must be connected down the outer surfaces of the cube. The result is that the chip-on-chip technique has typically been restricted to applications utilizing identical die with regular structures. For example, the most common application to date has been large memory devices constructed by stacking SRAM or DRAM die on top of each other.

A new technique which may serve to alleviate the problem of chip-on-chip interconnect is a process for creating vias through silicon substrates. Experiments are being performed in which aluminum "blobs" are placed on the surface of a silicon substrate and, by means of a gradient furnace, the aluminum migrates through the silicon providing vias from one side to the other (Figure 21.20).

Another technique more in keeping with the times is to create the vias by punching the aluminum through the silicon by means of a laser. These developments pave the way for double-sided silicon substrates with chips and interconnect on both sides. Additionally, they offer strong potential for interconnecting the die used in chip-on-chip structures.

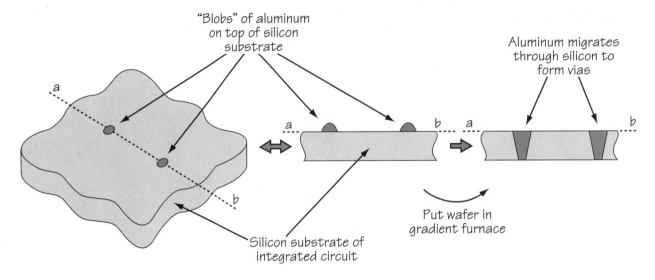

Figure 21.20: Chip-on-chip: creating vias through silicon

Figure 21.21: Conductive adhesives

Conductive Adhesives

Many electronics fabrication processes are exhibiting a trend towards mechanical simplicity with underlying sophistication in materials technology. A good example of this trend is illustrated by conductive, or *anisotropic,* adhesives which contain minute particles of conductive material.

These adhesives find particular application with the flipped-chip techniques used to mount bare die on the substrates of hybrids, multichip modules, or circuit boards. The adhesive is screen-printed onto the substrate at the site where the die is to be located, the die is pressed into the adhesive, and the adhesive is cured using a combination of temperature and pressure (Figure 21.21).

The beauty of this scheme is that the masks used to screen print the adhesive do not need to be too complex and the application of the adhesive does not need to be excessively precise, because it can be spread across all of the component pads. The conducting particles are only brought in contact with each other at the sites where the raised pads on the die meet their corresponding pads on the substrate, thereby forming good electrical connections.

The original conductive adhesives were based on particles such as silver. But, in addition to being expensive, metals like silver can cause electron migration problems at the points where they meet the silicon substrates. Modern equivalents are based on organic metallic particles, thereby reducing these problems.

In addition to being simpler and requiring fewer process steps than traditional methods, the conductive adhesive technique removes the need for solder, whose lead content is beginning to raise environmental concerns.

Superconductors

One of the "Holy Grails" of the electronics industry is to have access to conductors with zero resistance to the flow of electrons, and for such conductors, known as a *superconductors,* to operate at room temperatures. As a concept, superconductivity is relatively easy to understand: consider two sloping ramps into which a number of pegs are driven. In the case of the first ramp, the pegs are arranged randomly across the surface, while in the second the pegs are arranged in orderly lines. Now consider what happens when balls are released at the top of each surface:

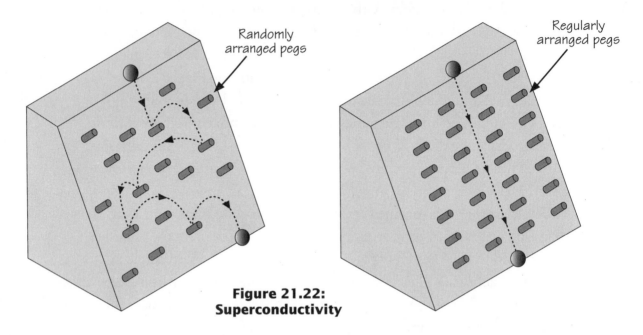

Figure 21.22:
Superconductivity

In the case of the randomly arranged pegs, the ball's progress is repeatedly interrupted, while in the case of the pegs arranged in orderly lines, the ball slips through like *"water off a duck's back."* Although analogies are always suspect (and this one doubly so), the ramps may be considered to represent conducting materials, the gravity accelerating the balls takes on the role of voltage differentials applied across the ends of the conductors, the balls play the part of electrons, and the pegs portray atoms.

The atoms in materials vibrate due to the thermal energy contained in the material: the higher the temperature, the more the atoms vibrate. An ordinary conductor's electrical resistance is caused by these atomic vibrations, which obstruct the movement of the electrons forming the current. Using the *Kelvin*,[26] or *absolute*, scale of temperature, 0 K (corresponding to –273°C) is the coldest possible temperature and is known as *absolute zero*. If an ordinary conductor were cooled to a temperature of

---

[26] Invented by the British mathematician and physicist William Thomas, first Baron of Kelvin.

absolute zero, atomic vibrations would cease, electrons could flow without obstruction, and electrical resistance would fall to zero. A temperature of absolute zero cannot be achieved in practice, but some materials exhibit superconducting characteristics at higher temperatures.[27]

In 1911, the Dutch physicist Heike Kamerlingh Onnes discovered superconductivity in mercury at a temperature of approximately 4K (–269°C). Many other superconducting metals and alloys were subsequently discovered but, until 1986, the highest temperature at which superconducting properties were achieved was around 23K (–250°C) with the niobium-germanium alloy (Nb_3Ge).

In 1986, Georg Bednorz and Alex Müller discovered a metal oxide that exhibited superconductivity at the relatively high temperature of 30K (–243°C). This led to the discovery of ceramic oxides that superconduct at even higher temperatures. In 1988, an oxide of thallium, calcium, barium and copper ($Tl_2Ca_2Ba_2Cu_3O_{10}$) displayed superconductivity at 125K (–148°C), and, in 1993, a family based on copper oxide and mercury attained superconductivity at 160K (–113°C). These *"high-temperature"* superconductors are all the more noteworthy because ceramics are usually extremely good insulators.

Like ceramics, most organic compounds are strong insulators; however, some organic materials known as organic synthetic metals do display both conductivity and superconductivity. In the early 1990s, one such compound was shown to superconduct at approximately 33K (–240°C). Although this is well below the temperatures achieved for ceramic oxides, organic superconductors are considered to have great potential for the future.

New ceramic and organic superconducting materials are being discovered on a regular basis, and the search is on for room temperature superconductors which, if discovered, are expected to revolutionize electronics as we know it.

---

[27] If the author were an expert in superconductivity, this is the point where he might be tempted to start muttering about *"Correlated electron movements in conducting planes separated by insulating layers of mesoscopic thickness, under which conditions the wave properties of electrons assert themselves and electrons behave like waves rather than particles."* But he's not, so he won't.

Nano-technology

Nano-technology is an elusive term that is used by different research-and-development teams to refer to whatever it is that they're working on at the time. However, regardless of their particular area of interest, nano-technology always refers to something extremely small; for example, motors and pumps the size of a pinhead, which are created using similar processes to those used to fabricate integrated circuits. In fact, around the beginning of 1994, one such team unveiled a miniature model car, which was smaller than a grain of short-grain rice. This model contained a micro-miniature electric motor, battery, and gear train, and was capable of traversing a fair-sized room (though presumably not on a shag-pile carpet).

One of the more outrageous branches of nano-technology that has been suggested as having potential in the future is that of micro-miniature electronic products that assemble themselves! The theory is based on the way in which biological systems operate. Specifically, the way in which *enzymes*[28] act as biological catalysts[29] to assemble large, complex molecules from smaller molecular building blocks.

Before commencing this discussion, it is necessary to return to the humble water molecule.[30] As you may recall, water molecules are formed from two hydrogen atoms and one oxygen atom, all of which share electrons between themselves. However, the electrons are not distributed equally, because the oxygen atom is a bigger, more robust fellow which grabs more than its fair share (Figure 21.23).

The angle formed between the two hydrogen atoms is 105°. This is because, of the six electrons that the oxygen atom owns, two are shared with the hydrogen atoms and four remain the exclusive property of the oxygen. These four huddle together on one side of the oxygen atom and put "pressure" on the bond angle. The bond angle settles on 105° because this is

---

[28] Complex proteins which are produced by living cells and catalyze biochemical reactions at body temperatures.

[29] A substance that initiates a chemical reaction under different conditions (such as lower temperatures) than would otherwise be possible. The catalyst itself remains unchanged at the end of the reaction.

[30] Water molecules were introduced in Chapter 2.

the point where the pressure from the four electrons is balanced by the natural repulsion of the two positively charged hydrogen atoms (similar charges repel each other).

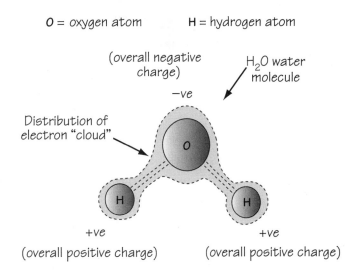

Figure 21.23: Nano-technology: distribution of electrons in a water molecule

The end result is that the oxygen atom has an overall negative charge, while the two hydrogen atoms are left feeling somewhat on the positive side. This unequal distribution of charge means that the hydrogen atoms are attracted to anything with a negative bias—for example, the oxygen atom of another water molecule. Although the strength of the resulting bond, known as a *hydrogen bond*, is weaker than the bond between the hydrogen atom and its "parent" oxygen atom, it is still quite respectable.

When water is cooled until it freezes, its resulting crystalline structure is based on these hydrogen bonds. Even in its liquid state, the promiscuous, randomly wandering water molecules are constantly forming hydrogen bonds with each other. These bonds persist for a short time until another water molecule clumsily barges into them and knocks them apart. From this perspective, a glass of water actually contains billions of tiny ice crystals that are constantly forming and being broken apart again.

However, we digress. Larger molecules can form similar electrostatic bonds with each other. Imagine a "soup" consisting of large quantities of many different types of molecules, two of which, M_a and M_b, may be combined to form larger molecules of type M_{ab} (Figure 21.24).

This is similar in concept to two pieces of a jigsaw, which will only fit together if they are in the correct orientation to each other. Similarly, M_a and M_b will only bond to form M_{ab} if they are formally presented to each other in precisely the right orientation. However, the surfaces of the molecules are extremely complex three-dimensional shapes, and achieving the

There are several theories as to how life developed on Earth, including "seeding" by meteors from outer space[31] or random combinations of atoms forming simple molecules, which in turn formed rudimentary proteins, then . . . and so it goes.

Maybe a combination of both of these theories is true. However, to test the *"random combination"* theory, scientists decided to replicate the conditions on Earth hundreds of millions of years ago before life developed, with the intention of duplicating the happy event. So, they took a large glass jar, added some water, filled the rest with the gases that were around at the time, and applied electrical discharges to the mixture to simulate the effects of lightning strikes. Much to their surprise . . . nothing happened.

Obviously this was something of a downer. While they were all scratching their heads, a tentative hand was raised in the air, and a high-pitched voice belonging to the tea boy said "Excuuussse me, but what about adding rocks?" After they had slapped him round the back of the head for his insolence, they thought "What the heck, let's give it a whirl." So, they added a mixture of rocks, zapped everything with electric discharges again, and, lo and behold, they found amino acids and simple proteins.

It turned out that when certain shales are split, their rough surfaces form exactly the right templates to bond atoms and molecules together to create the rudimentary building blocks of life!

Now, this does not either prove or disprove the existence of God. However, there is almost certainly something somewhere, because everything fits together so perfectly. At the very least, if there is a God, then it certainly serves to confirm the fact that *"He/She works in mysterious ways, His/Her wonders to perform."*

---

[31] Scientists have found amino acids (the basic building blocks of proteins, and therefore life itself) in a number of meteors.

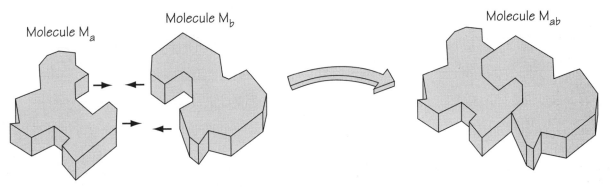

**Figure 21.24: Nano-technology:
combining molecules M$_a$ and M$_b$ to form M$_{ab}$**

correct orientation is a tricky affair. Once the molecules have been brought together their resulting bonds are surprisingly strong, but the chances of the two molecules randomly achieving exactly the correct orientation to form the bonds are extremely small.

It is at this point of the story that enzymes reenter the plot. There are numerous enzymes, each dedicated to the task of "matchmaking" for two of their favorite molecules. The surface of an enzyme is also an extremely complex three-dimensional shape, but it is much larger than its target molecules and has a better chance of gathering them up. The enzyme floats around until it bumps into a molecule of type M$_a$ to which it bonds. The enzyme then continues on its trek until it locates a molecule of type M$_b$. When the enzyme bonds to molecule M$_b$, it orientates it in exactly the right way to complete the puzzle with molecule M$_a$ (Figure 21.25).

The bonds between M$_a$ and M$_b$ are far stronger than their bonds to the enzyme. In fact, as soon a these bonds are formed, the enzyme is actually repelled by the two little lovebirds and promptly thrusts M$_{ab}$ away. However, the enzyme immediately forgets its pique, and commences to search for two more molecules (some enzymes can catalyze their reactions at the rate of half a million molecules per minute).

The saga continues, because another, larger enzyme may see its task in life as bringing M$_{ab}$ together with yet another molecule M$_{cd}$. And so it continues, onwards and upwards, until the final result, whatever that may be, is achieved.

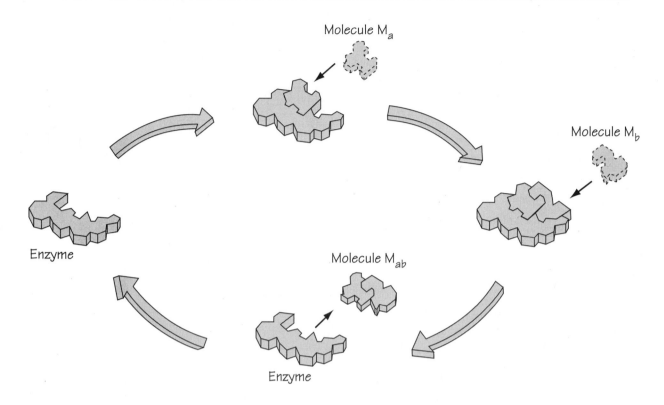

Molecule M$_a$

Molecule M$_b$

Enzyme

Molecule M$_{ab}$

Enzyme

**Figure 21.25: Nano-technology:
forming molecule M$_{ab}$ using an enzyme**

As our ability to create *"designer molecules"* increases, it becomes increasingly probable that we will one day be able to create *"designer enzymes."* This would enable us to mass-produce *"designer proteins"* that could act as alternatives to semiconductors (see also the *"Protein Switches and Memories"* section above). As one of the first steps along this path, a process could be developed to manufacture various proteins that could then be bonded to a substrate or formed into three-dimensional blocks for optical memory applications. At a more sophisticated level, it may be possible for such a process to directly create the requisite combinations of proteins as self-replicating structures across the face of a substrate.

However, the possibilities extend far beyond the mass-production of proteins. It is conceivable that similar techniques could be used to assemble nonorganic structures such as microscopic electromechanical artifacts. All that would be required (he said casually) would be for the individual components to be shaped in such a way that naturally occurring electro-static fields would cause them to form bonds when they were brought together with their soul mates. In fact, this is one step along the path towards molecular-sized robots known as *nanobots*. Taken to extremes, the discipline of electronics in the future may not involve the extreme temperatures, pressures, and noxious chemicals that are in vogue today. Instead, electronics may simply involve "cookbook" style recipes; for example, the notes accompanying an electronics course in 2050 AD may well read as follows:

Intermediate Electronics (Ages 12 to 14)
Super Computers 101

Instructions for creating a micro-miniature massively parallel super-computer.*

   a) Obtain a large barrel.

   b) In your barrel, mix two parts water and one part each of chemicals A, B, C . . .

   c) Add a pinch of nanobot-mix (which you previously created in Nanobots 101).

   d) Stir briskly for one hour with a large wooden spoon.

Congratulations, you will find your new super-computers in the sediment at the bottom of the barrel. Please keep one teaspoon of these super computers for your next lesson.

* These instructions were reprooduced from "*Bebop to the Boolean Boogie,*" 50th Edition, "The most successful electronics book in the history of the universe"—over 300,000,000 copies sold, by kind permission of HighText Publications Inc.

Of course, some of this is a little far-fetched (with the hopeful exception of the references to *Bebop to the Boolean Boogie*). However, for what it's worth, the author would bet his wife's life savings that this type of technology will occur one day, and also that it will be here sooner than you think!

Summary

The potpourri of technologies introduced above have been offered for your delectation and delight. Some of these concepts may appear to be a little on the wild side, and you certainly should not believe everything that you read or hear. On the other hand, you should also be careful not to close your mind, even to seemingly wild and wacky ideas, in case something sneaks up behind you and bites you on the ****.[32] As the Prize said in *Where is Earth* by Robert Sheckley: *"Be admiring but avoid the fulsome, take exception to what you don't like, but don't be stubbornly critical; in short, exercise moderation except where a more extreme attitude is clearly called for."*

And so, with our lower lips quivering and little tears rolling down our cheeks, we come to the close of this, the final chapter. But turn that frown upside down into a smile . . . there are still oodles of yummy appendices containing enough mouth-watering information to form a book in their own right. As that great British Prime Minister Winston Spencer Churchill (1874–1965) would have said: *"Now this is not the end. It is not even the beginning of the end. But it is, perhaps, the end of the beginning."*[33]

Light waves become distorted and attenuated as they travel through optical fibers—similar to the way in which electrical pulses deteriorate as they propagate through conducting wires. However, there are certain forms of waves called solitons that, due to their contour, can retain their shape and speed. One example of a soliton is the Severn bore, a wave caused by unusual tidal conditions (where the Severn is a river that separates England and Wales). This bore (wave) can travel miles up the river without significant diminishment or deformation.

Similarly, scientists and engineers are experimenting with light solitons—pulses that can travel long distances through optical fibers without distortion. Using light solitons, data can retain its integrity while being transmitted over 20,000 kilometers through optical fibers at a rate of 10 gigabits per second—four times faster than is possible using conventional techniques.

---

[32] Arsek no questions!

[33] Speech at the Lord Mayor's Day Luncheon, London, England (November 10, 1942).

Assertion-Level Logic

The purpose of a circuit diagram is to convey the maximum amount of information in the most efficient fashion. One aspect of this is assigning meaningful names to wires; for example, naming a wire `system_reset` conveys substantially more information than calling it `big_boy`. Another consideration is that the signals carried by wires may be active-high or active-low (see sidebar). Consider a portion of a circuit containing a tri-state buffer with an active-low enable (Figure A.1).

Figure A.1: Naming active-low signals

Beware—here be dragons!

The definitions of active-high and active-low are subject to confusion. Some academics (and even text books) define an active-low signal as one whose asserted (TRUE or logic 1) state is at a lower voltage level than its unasserted (FALSE or logic 0) state. However, this definition serves only to cause confusion when combined with negative-logic implementations (see also Appendix B).

Engineers in the trenches generally take the position that an active-high signal is one whose active state is considered to be TRUE or logic 1, while an active-low signal is one whose active state is considered to be FALSE or logic 0. These definitions allow all forms of logic, including assertion-level logic, to be represented without any confusion, regardless of whether positive- or negative-logic implementations are employed. These manly-man definitions are the ones used throughout this book. However, when using the terms active-high and active-low in discussions with other people, you are strongly advised to make sure that you're all talking about the same thing before you start.

Instead of a tilda, some designers prefer to use an exclamation mark '!' nicknamed a *shriek* to indicate active-low signals—for example, !enable. Alternatively, designers may prefix or postfix active-low signal names with a special letter; for example, Nenable or enableB. Yet another technique is to draw a horizontal line, or bar, over the name, but this is not recommended for a number of reasons. First, signal names with bars over them are difficult to replicate in textual form on a computer. Second, bars are often used to indicate negations in Boolean equations. In this latter case it would be impossible to determine whether a bar formed an integral part of a signal's name or if it was intended to indicate that the signal should be negated. Tilda characters may also be used to indicate complementary outputs; for example, the q and ~q outputs from a latch or a flip-flop.

To convey as much information as possible, it is preferable to indicate the nature of an active-low signal in its name. One method of achieving this is to prefix the name with a tilda character '~'; hence, ~enable (see sidebar). When both en_a and en_b are set to their active-high states, ~enable is driven to its active-low state and the tri-state buffer is enabled. However, it is important to note that the active-low nature of ~enable is not decided by the NAND. The only thing that determines whether a signal is active-high or active-low is the way in which it is used by any target gates such as the tri-state buffer in this example. Thus, active-low signals can be generated by ANDs and ORs as easily as NANDs and NORs.

The problem with the standard symbols for BUF, NOT, AND, NAND, OR, and NOR is that they are tailored to reflect operations based on active-high signals being presented to their inputs. To address this problem, special assertion-level logic symbols can be used to more precisely indicate the function of gates with active-low inputs. Consider a NOT used to invert the ~enable signal:

Figure A.2: Standard versus assertion-level NOT symbols

Both symbols indicate an inversion, but the assertion-level symbol is more intuitive because it reflects the fact that an active-low input is being transformed into an active-high output. In both cases the bobbles on the symbols indicate the act of inversion; one way to visualize this is that the symbol for a NOT has been forced into a symbol for a BUF until only the bobble remains visible (Figure A.3).

In the real world, both standard and assertion-level symbols are implemented using identical logic gates. Assertion-level logic does not affect the final implementation, but simply offers an alternative way of viewing things. Visualizing bobbles as representing inverters is a useful technique for handling more complex functions. Consider a variation on our original circuit containing a tri-state buffer with an active-high enable (Figure A.4).

The purpose of this circuit is to set the *enable* signal to its active-high state if either ~*en_a* or ~*en_b* are in their active-low states. Once again, the assertion-level representation

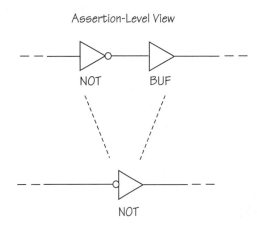

Figure A.3: Implications of standard versus assertion-level NOT symbols

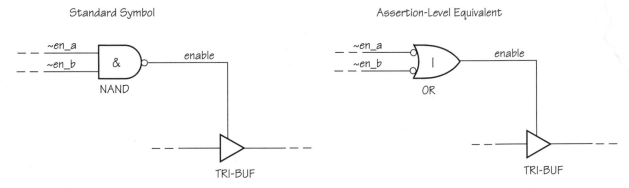

Figure A.4: Standard NAND versus assertion-level OR symbols

is the more intuitive, especially for someone who is unfamiliar with the function of the circuit; the assertion-level symbol unambiguously indicates that *enable* will be set to logic 1 if either *~en_a* or *~en_b* are presented with logic 0s.

Any standard primitive gate symbol can be transformed into its assertion-level equivalent by inverting all of its inputs and outputs, and exchanging & operators for | operators and vice versa. In fact, a quick review of Chapter 9 reveals that these steps are identical to those used in a DeMorgan Transformation. Thus, assertion-level symbols may also be referred to as DeMorgan equivalent symbols. The most commonly used assertion-level symbols are those for BUF, NOT, AND, NAND, OR, and NOR (Figure A.5).

Note that the assertion-level symbols for XOR and XNOR are identical to their standard counterparts. The reason for this is that if you take an XOR and invert its inputs and outputs, you end up with an XOR again; similarly for an XNOR. A little experimentation with their truth tables will quickly reveal the reason why.

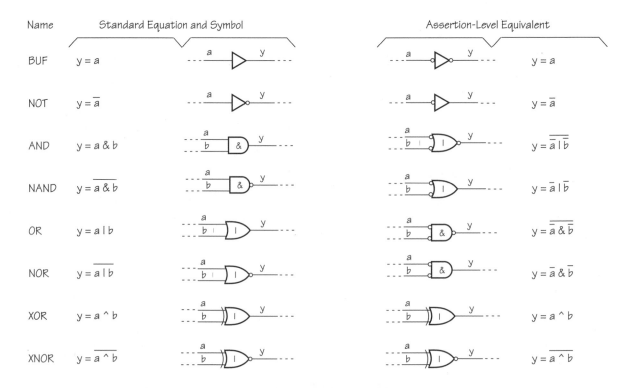

Figure A.5: Commonly used assertion-level symbols

Positive Logic versus Negative Logic

The terms positive logic and negative logic refer to conventions which dictate the relationship between logical values and the physical voltages used to represent them. Unfortunately, although the core concepts are relatively simple, fully comprehending all the implications requires an exercise in lateral thinking sufficient to make even a strong man break down and weep!

Before plunging into these conventions, it is important to understand that logic 0 and logic 1 are *always* equivalent to FALSE and TRUE, respectively.[1] The reason that these terms are used interchangeably is that digital functions can be considered to represent either logical or arithmetic operations (Figure B.1).

It is generally preferable to employ a single consistent format to cover both cases, and it is easier to view logical operations in terms of 0s and 1s than to view arithmetic operations in terms of Fs and Ts. The point here is that, by definition, logic 0 and logic 1 are logical concepts which have no direct relationship to any physical values.

a	b	y
F	F	F
F	T	T
T	F	T
T	T	F

XOR

LOGICAL OPERATION

a	b	y
0	0	0
0	1	1
1	0	1
1	1	0

XOR

ARITHMETIC OPERATION
(MODULO-2 ADDER)

Figure B.1: Logical versus arithmetic views of digital functions

---

[1] Unless you're really taking a walk on the wild side, in which case all bets are off.

Physical to Abstract Mapping (NMOS Logic)

The process of relating logical values to physical voltages begins by defining the frames of reference to be used. One absolute frame of reference is provided by truth tables which are always associated with specific functions:

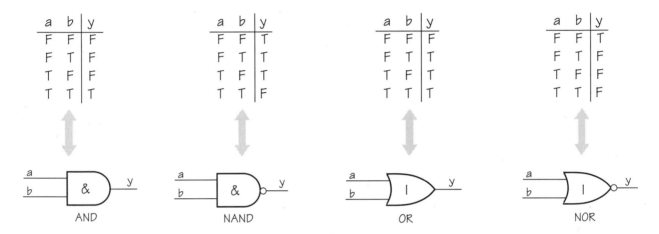

a	b	y
F	F	F
F	T	F
T	F	F
T	T	T

AND

a	b	y
F	F	T
F	T	T
T	F	T
T	T	F

NAND

a	b	y
F	F	F
F	T	T
T	F	T
T	T	T

OR

a	b	y
F	F	T
F	T	F
T	F	F
T	T	F

NOR

Figure B.2: Absolute relationship between truth tables and functions

Another absolute frame of reference is found in the physical world; specific voltage levels applied to the inputs of a digital function cause corresponding voltage responses on the outputs. These relationships can also be represented in truth table form; consider a logic gate constructed using only NMOS transistors:

Physical Mapping

a	b	y
0v	0v	+ve
0v	+ve	+ve
+ve	0v	+ve
+ve	+ve	0v

Figure B.3: Physical mapping (NMOS logic gate)

With NMOS transistors connected as shown above, an input connected to the more negative V_{SS} turns that transistor OFF, and an input connected to the more positive V_{DD} turns that transistor ON. The final step is to define the mapping between the physical and abstract worlds; either Ov is mapped to FALSE and +ve is mapped to TRUE or vice versa:

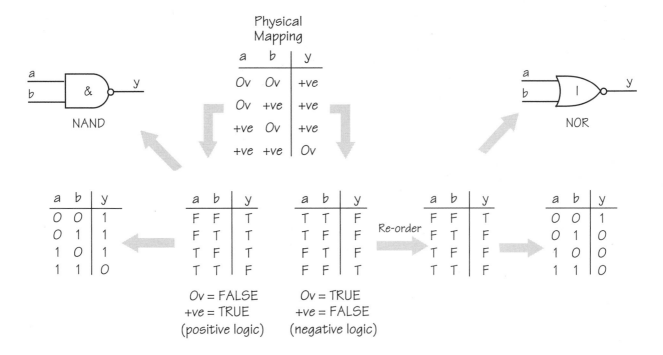

Figure B4: Physical to abstract mapping (NMOS logic gate)

Using the positive logic convention, the more positive potential is considered to represent TRUE[2] and the more negative potential is considered to represent FALSE. By comparison, using the negative logic convention, the more negative potential is considered to represent TRUE[3] and the more positive potential is considered to represent FALSE. Thus, this circuit may be considered to be performing either a NAND function in positive logic or a NOR function in negative logic.

---

[2] Hence, positive logic is also known as positive-true.

[3] Hence, negative logic is also known as negative-true.

Physical to Abstract Mapping (PMOS Logic)

From the previous example it would appear that positive logic is the more intuitive as it is easy to relate logic 0 to $0v$ (no volts) and logic 1 to $+ve$ (presence of volts). On this basis one may wonder why negative logic was ever invented. The answer to this is, as in so many things, rooted in history. When the MOS technology was originally developed, PMOS transistors were easier to manufacture and were more reliable than their NMOS counterparts. Thus, the majority of early MOS logic gates were constructed from combinations of PMOS transistors and resistors. Consider a logic gate constructed using only PMOS transistors:

a	b	y
$0v$	$0v$	$-ve$
$0v$	$-ve$	$-ve$
$-ve$	$0v$	$-ve$
$-ve$	$-ve$	$0v$

Figure B.5: Physical mapping (PMOS logic gate)

The original PMOS logic typically made use of a negative power supply. With PMOS transistors connected as shown above, an input connected to the more positive V_{SS} turns that transistor OFF, and an input connected to the more negative potential V_{DD} turns that transistor ON. Once again, the final step is to define the mapping between the physical and abstract worlds; either $0v$ is mapped to FALSE and $-ve$ is mapped to TRUE or vice versa (Figure B.6).

Thus, this circuit may be considered to be performing either a NOR function in positive logic or a NAND function in negative logic. In this case, negative logic is the more intuitive as it is easy to relate logic 0 to $0v$ (no volts) and logic 1 to $-ve$ (presence of volts). Additionally, the physical

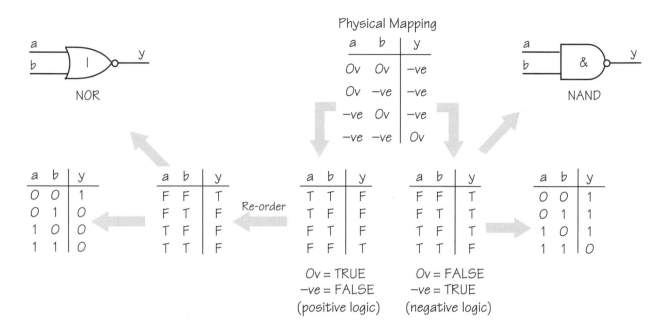

**Figure B.6: Physical to abstract mapping
(PMOS logic gate)**

structure of the PMOS gate is identical to that of the NMOS gate; if the NMOS gate is represented in positive logic and the PMOS gate is represented in negative logic, then both representations equate to NAND functions which is, if nothing else, aesthetically pleasing.

Physical to Intermediate to Abstract Mapping

The discussions above offer the most consistent way to view the positive and negative logic conventions; the casual reader is advised to stop at this point and proceed to something less demanding on the brain.[4] Occasionally, an additional level of indirection is included in the progression from the physical to the abstract worlds. One reason for this is that supply and switching voltage values are relative rather than absolute quantities. In the examples above, the NMOS gate is shown as being powered by $V_{SS} = 0v$ and $V_{DD} = +ve$, while the PMOS gate is shown as being powered by $V_{SS} = 0v$ and $V_{DD} = -ve$.

---

[4] You have been warned!

However, within certain physical constraints, the actual values of V_{DD} and V_{SS} are largely immaterial. The only real requirements are for V_{DD} to be more positive than V_{SS} for NMOS gates, and for V_{DD} to be more negative than V_{SS} for PMOS gates. Assuming that both technologies require a 5 volt differential between V_{DD} and V_{SS}, the following shows a small selection of the many possible voltage combinations that would suffice:

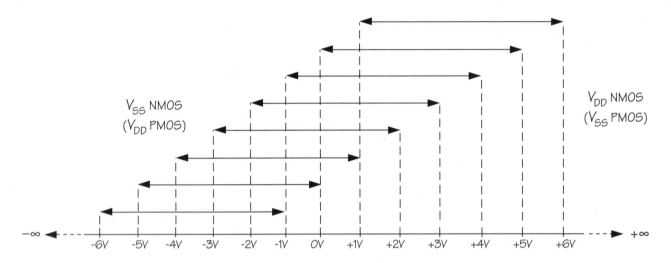

Figure B.7: Alternative values for V_{DD} and V_{SS}

Now consider an NMOS gate using V_{DD} = +2 volts and V_{SS} = −3 volts. Although it is preferable to map from the physical world directly to FALSE and TRUE values, it is also possible to map to intermediate HIGH and LOW values. The problem arises in the definitions of HIGH and LOW; some would regard HIGH as being the more positive potential and LOW as being the more negative, while others would adopt absolute values as measured from 0 volts as being the criteria. In this latter case, −3 volts would be considered to be HIGH as it has a greater absolute value with respect to 0 volts than +2 volts which would therefore be taken to be LOW.

Therefore, for an NMOS gate supplied by V_{DD} = +2v and V_{SS} = −3v, the progression of mapping from physical through intermediate to abstract representations could be as shown in Figure B.8. It is easy to see how this progression can become extremely confusing. To further illustrate the

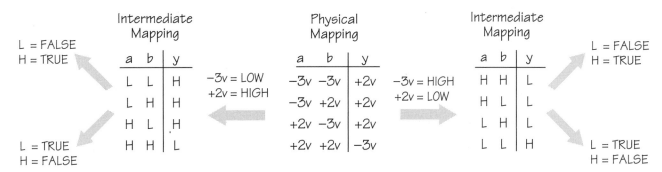

Figure B.8: Physical to intermediate to abstract mapping (NMOS logic gate)

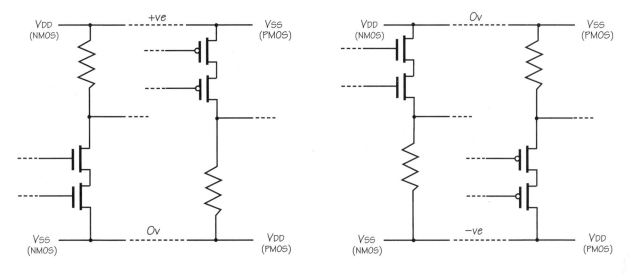

Figure B.9: NMOS and PMOS logic gates using common voltage levels

possible complexities involved, consider a design containing a combination of NMOS and PMOS logic gates.[5] Employing three voltage levels, *0v*, *+ve* and *−ve*, would be less than satisfactory because the gates would not be able to communicate with each other. However, there are solutions requiring only two voltage levels (Figure B.9).

---

[5] Note that this is not the same as using NMOS and PMOS transistors to form CMOS gates.

As was noted above, the only real requirements are for V_{DD} to be more positive than V_{SS} for NMOS gates, and for V_{DD} to be more negative than V_{SS} for PMOS gates. Both these requirements can be satisfied using only two voltage levels as long as the labels associated with those levels are redefined as necessary. The two examples above are actually different views of the same thing; any two voltage levels with the appropriate relationship to each other would suffice. The application of the positive or negative logic systems in a mixed scenario of this type would prove to be an interesting exercise.

Fortunately, the majority of modern technologies favor positive logic representations and, unless otherwise stated, the positive logic convention is typically assumed. However, it is ultimately the designers' responsibility to define and/or ascertain which convention is being employed in any particular design.

Reed-Müller Logic

Some digital functions can be difficult to optimize if they are represented in the conventional sum-of-products or product-of-sums forms which are based on ANDs, ORs, NANDs, NORs, and NOTs. In certain cases it may be more appropriate to implement a function in a form known as Reed-Müller logic which is based on XORs and XNORs. One indication as to whether a function is suitable for the Reed-Müller form of implementation is if that function's Karnaugh map displays a checkerboard pattern of 0s and 1s. Consider a familiar two-input function:

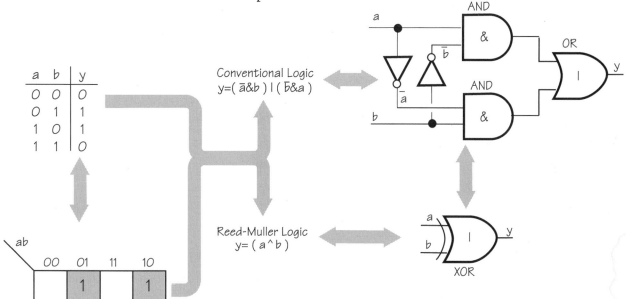

Figure C.1: Two-input function suitable for Reed-Müller implementation

Since the above truth table is easily recognizable as being that for an XOR function, it comes as no great surprise to find that implementing it as a single XOR gate is preferable to an implementation based on multiple AND, OR and NOT gates. A similar checkerboard pattern may apply to a three-input function (Figure C.2).

Figure C.2: Three-input function suitable for Reed-Müller implementation

As XORs are both commutative[1] and associative,[2] it does not matter which combinations of inputs are applied to the individual gates. The checkerboard pattern for a four-input function continues the theme (Figure C.3).

Larger checkerboard patterns involving groups of 0s and 1s also indicate functions suitable for Reed-Müller implementation (Figure C.4).

Having recognized a checkerboard pattern, there is a quick "*rule of thumb*" for determining the variables to be used in the Reed-Müller implementation. Select any group of 0s or 1s and identify the significant and redundant variables,[3] and then simply XOR the significant variables together.

---

[1] For example, $(a \wedge b) \equiv (b \wedge a)$.

[2] For example, $(a \wedge b) \wedge c \equiv a \wedge (b \wedge c)$.

[3] The significant variables are those whose values are the same for all the boxes forming the group, while the redundant variables are those whose values vary between boxes.

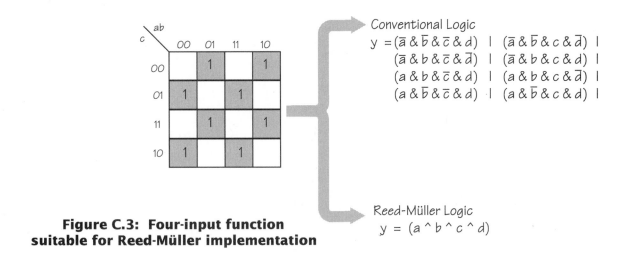

Conventional Logic
$$y = (\overline{a} \& \overline{b} \& \overline{c} \& d) \mid (\overline{a} \& \overline{b} \& c \& \overline{d}) \mid$$
$$(\overline{a} \& b \& \overline{c} \& \overline{d}) \mid (\overline{a} \& b \& c \& d) \mid$$
$$(a \& b \& \overline{c} \& d) \mid (a \& b \& c \& \overline{d}) \mid$$
$$(a \& \overline{b} \& \overline{c} \& d) \mid (a \& \overline{b} \& c \& d) \mid$$

Reed-Müller Logic
$$y = (a \wedge b \wedge c \wedge d)$$

**Figure C.3: Four-input function
suitable for Reed-Müller implementation**

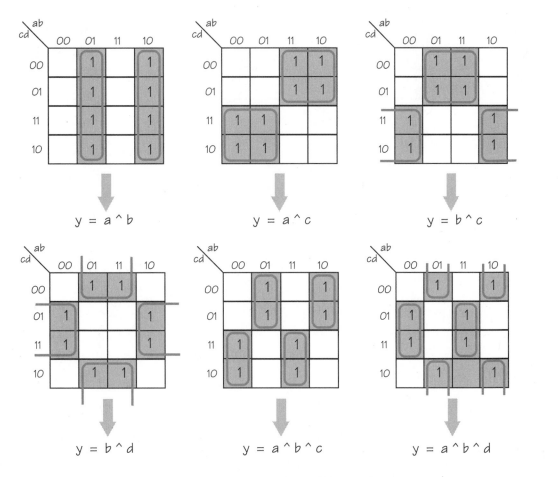

$$y = a \wedge b \qquad\qquad y = a \wedge c \qquad\qquad y = b \wedge c$$

$$y = b \wedge d \qquad\qquad y = a \wedge b \wedge c \qquad\qquad y = a \wedge b \wedge d$$

**Figure C.4: Additional functions suitable
for Reed-Müller implementation**

As all of the checkerboard patterns above include a logic 0 in the box in the upper left corner,[4] the resulting Reed-Müller implementations can be realized using only XORs. However, any pair of XORs may be replaced with XNORs, the only requirement being that there are an even number of XNORs.

If the checkerboard pattern includes a logic 1 in the box in the upper left corner, the Reed-Müller implementation must contain an odd number of XNORs. Once again, it does not matter which combinations of inputs are applied to the individual XORs and XNORs:

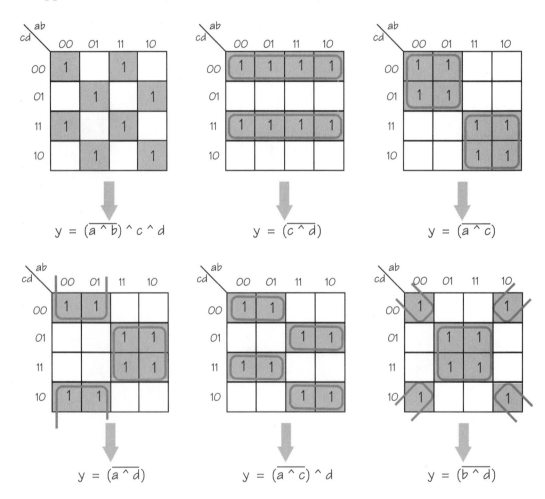

Figure C.5: Reed-Müller implementations requiring XNOR gates

---

[4] Corresponding to all of the inputs being logic 0.

Although these examples provide a very limited introduction to the concept of Reed-Müller logic, checkerboard Karnaugh map patterns are easy to recognize and appear surprisingly often. Reed-Müller implementations are often appropriate for circuits performing arithmetic or encoding functions.[5]

---

[5] See also Appendix D.

Gray Codes

W hen moving between states in a standard binary sequence, multiple bits may change from 0 to 1 or vice versa; for example, three bits change value when moving from 0011_2 to 0100_2. In the physical world there is no way to ensure that all the bits will transition at exactly the same time, so a system may actually pass through a sequence of intermediate states. An intended state change of 0011_2 to 0100_2 could actually result in the sequence 0011_2 to 0111_2 to 0100_2.

One way to avoid this problem is to use a Gray code in which only a single bit changes when moving between states:

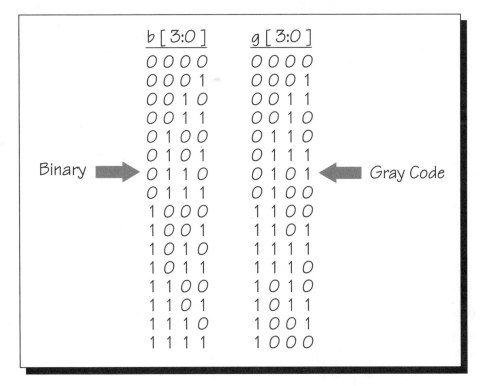

Figure D.1: Binary versus Gray code

Gray codes are of use for a variety of applications such as the ordering of the input variables on Karnaugh maps.[1] Another application is encoding the angle of a mechanical shaft, where a disc attached to the shaft is patterned with areas of conducting material:

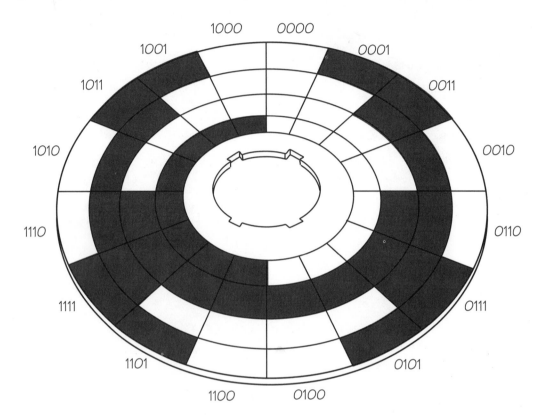

Figure D.2: Gray code used for shaft-angle encoding

The conducting areas are arranged in concentric circles where each circle represents a binary digit. A set of electrical contacts, one for each circle, are used to detect the logic values represented by the presence or absence of the conducting areas. A digital controller can use this information to determine the angle of the shaft. The precision of the measurement depends on the number of bits, or circles. A 4-bit value representing 16 unique states allows the shaft's angle to be resolved to 22.5 degrees, while a 10-bit value representing 1024 unique states allows the shaft's angle to be

_____

[1] This was discussed in detail in Chapter 10.

resolved to 0.35 degrees. Using a Gray code sequence to define the conducting and non-conducting areas ensures that no intermediate values are generated as the shaft rotates.

Commencing with a state of all zeros, a Gray code can be generated by always changing the least significant bit that results in a new state. An alternative method which may be easier to remember and use is as follows:

a) Commence with the simplest Gray code possible; that is, for a single bit.

b) Create a mirror image of the existing Gray code below the original values.

c) Prefix the original values with 0s and the mirrored values with 1s.

d) Repeat steps b) and c) until the desired width is achieved.

An example of this mirroring process used to generate a 4-bit Gray code would be as follows:

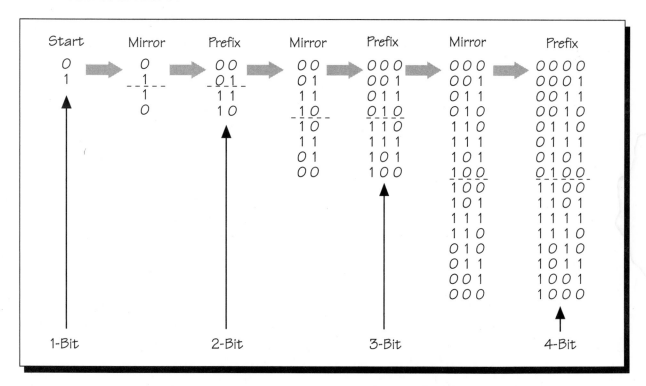

Figure D.3: Generating a 4-bit Gray code using the mirroring process

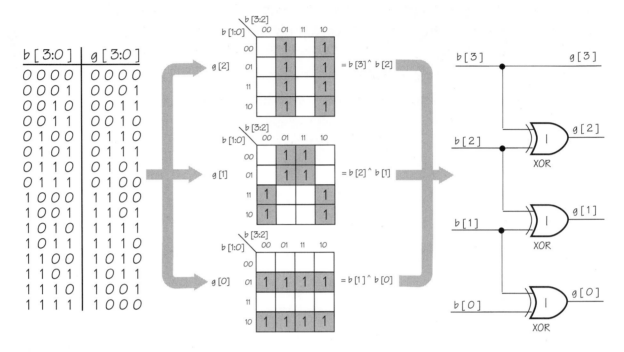

Figure D.4: Binary to Gray code converter

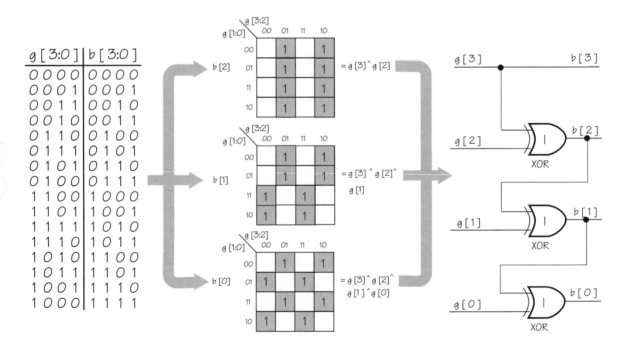

Figure D.5: Gray code to binary converter

It is often required to convert a binary sequence into a Gray code or vice versa. Such converters are easy to create and are primarily of interest here due to their affinity to Reed-Müller implementations.[2] Consider a binary to Gray code converter (Figure D.4). The checkerboard patterns of 0s and 1s in the Karnaugh maps immediately indicate the potential for Reed-Müller implementations.

Similar checkerboard patterns are also seen in the case of a Gray code to binary converter (Figure D.5).

---

[2] Introduced in Appendix C.

A Reed-Müller Extraction Utility

In this appendix we discover a thought-provoking little mind game to delight the secret computer programmer that lurks within all of us.[1,2,3] After reading the previous two appendices, you should be a "master of the mystic arts" of Reed-Müller logic and Gray codes. This is rather fortunate because, without this knowledge, what you are about to read would be likely to make your brains leak out of your ears. The simple program discussed below contains the never-before-seen-in-public, soon-to-be-famous, mind-bogglingly-convoluted, *Max's Algorithm* for automatically extracting Reed-Müller expressions.

Before examining the program which, by some strange quirk of fate, is called *max_prog*, it would be useful to know what it actually does. First, you create a text file representation of one or more truth tables that you would like to examine and, second, you run the program using this file as its input. In return, the program determines which truth tables are suitable for Reed-Müller implementations and returns the appropriate expressions. For example, consider the truth tables in Figure E.1.

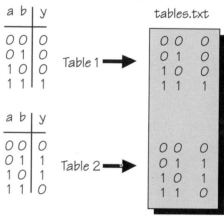

Figure E.1: Example truth tables for Reed-Müller evaluation

---

[1] The program is written in the C programming language, and is intended to be compatible with any ANSI-standard C compiler.

[2] In certain cases, good programming practices have been thrown out of the window for the sake of brevity. For example, functions are generally not declared before they are called, and no error checking is performed on the input data.

[3] If you'd like to know more about writing programs in C, you can't do much better than reading *The C Programming Language* (Second Edition) by Brian W. Kernighan and Dennis M. Ritchie. For those of you who wish to voyage a little further, *Reusable Data Structures For C* by Roger Sessions is well worth perusing.

As you will see, the part of the program that reads the input file is rather simple, hence the file's somewhat minimalist look. When you've created the input file, you run the program from the operating system's command line as follows:

```
max_prog < tables.txt                              { Calls the program, 'max_prog', and    }
                                                   { feeds it with the text file, 'tables.txt'. }


Truth table 1: No Reed-Muller solution            { These are the results returned by the }
Truth table 2: Reed-Muller solution is y = (a ^ b) { program for the tables shown above.  }
```

As truth tables 1 and 2 are obviously those for an AND and an XOR respectively, the results returned by the program should come as no surprise. Note that this version of the program assumes that the inputs to the truth tables are named a, b, c, … etc., and that the output is named y. In fact, this version of the program can handle truth tables with up to ten inputs (but only one output), and the number of inputs could be easily extended if necessary. Now, on to the program itself, the first part of which simply declares anything that the program is going to use:[4]

```
/* Max's algorithm to detect and evaluate functions suitable for Reed-Muller evaluation.  */
/* Accepts truth tables with up to ten inputs and returns any Reed-Muller equations.       */
/* Beware! Contains minimal parsing and no error detection - you have been warned!         */

#include <stdio.h>           /* Include the standard files supplied with C.                */
#include <ctype.h>
#include <string.h>

#define XSIZE 1024           /* Set the maximum number of elements necessary               */
                            /* to handle truth tables with two to ten inputs.             */

int bin_to_gray[XSIZE];      /* This will be the binary to Gray code lookup table.         */
int gray_to_bin[XSIZE];      /* This will be the Gray code to binary lookup table.         */

char workspace[XSIZE];       /* This will be the working area. Each element                */
                            /* corresponds to a box in a Karnaugh map.                    */
```

---

[4] If you do decide to play with this program, it is necessary to enter all the program segments in the same order that they are presented here.

```
struct xy {
  int x, y;
} xy_table[11] = {            /* When presented with the number of inputs      */
    { 0, 0},  /* 0 inputs */  /* associated with a truth table, this structure will */
    { 2, 1 }, /* 1 inputs */  /* return the most appropriate dimensions for an  */
    { 2, 2 }, /* 2 inputs */  /* equivalent Karnaugh map (as nearly rectangular */
    { 4, 2 }, /* 3 inputs */  /* as possible).                                  */
    { 4, 4 }, /* 4 inputs */  /*                                                */
    { 8, 4 }, /* 5 inputs */  /* The 'x' and 'y' values returned by the structure */
    { 8, 8 }, /* 6 inputs */  /* represent the number of columns (across) the   */
    {16, 8 }, /* 7 inputs */  /* map and the number of rows (down) the map      */
    {16,16 }, /* 8 inputs */  /* respectively.                                  */
    {32,16 }, /* 9 inputs */
    {32,32 } /*10 inputs */
};
```

We will return to discuss the lookup tables and the working area in a little while, but first let us move briskly on to the function that controls everything:

```
int main()
{
  int  num_tt = 0,        /* The number of the truth table currently being processed. */
       num_in,            /* The number of inputs in the current truth table.  */
       mask;              /* A binary mask which is used to track the significant inputs. */

  load_lookup_tables();                        /* Loads the lookup tables.     */

  while (num_in = get_truth_tables()) {         /* Load one truth table at a time. */
    num_tt++;                                   /* Keep track of the No. of the table. */
    if (test_reed_muller(num_in)) {             /* If this table has a R-M solution ... */
      mask = extract_reed_muller(num_in);       /* ... Extract the expression.  */
      output_reed_muller(mask,num_tt,num_in);   /* ... Output the equation.     */
    }
    else                                        /* If no R-M solution then say so. */
      printf("Truth table %i: No Reed-Muller solution\n",num_tt);
  }                                             /* Repeat for all the tables until none */
  return 0;                                     /* remain, then return 0 and exit. */
}
```

The variable num_tt which keeps track of the number of the truth table being processed is initialized with zero, then the function load_lookup_tables()

is called to initialize the lookup tables (what a surprise). The program then enters a loop as follows.

The function get_truth_tables() loads the outputs from the first truth table into the working area. This function also returns the number of inputs for the truth table and assigns the value to the variable num_in. When get_truth_tables() returns a value of zero for the number of inputs, then all of the truth tables in the input file have been processed and the program terminates. Otherwise, the first action in the loop is to increment num_tt, which keeps track of the number of the truth table that is being analyzed.

Once a truth table has been loaded, the function test_reed_muller() examines it to determine if it has a Reed-Müller solution. If there is such a solution, then the function extract_reed_muller() is used to determine which of the truth table's inputs are significant. The information on the significant inputs is presented in the form of a bit-mask which is assigned to the variable mask. Finally, the function output_reed_muller() displays the Reed-Müller equation for all the world to see. Good grief, if only everything in life could be so simple. Now, before plunging into the inner workings of the program, there are two utility functions that are called from time to time:

```
/* power_two      : Generates a power of two.                                       */
/*                 : Accepts an integer 'exponent'.                                 */
/*                 : Returns an integer equal to 2 raised to the power of 'exponent'. */
int power_two(int exponent)
{
   return 1 << exponent;         /* Returns 1 shifted to the left by the number of bits defined */
}                                /* by 'exponent' (which equals 2 to the power of 'exponent').  */
```

```
/* get_binary_value : Extracts a binary value presented as a string of 0 and 1 characters. */
/*                  : Accepts a pointer to the string of characters, '*p'.                 */
/*                  : Returns the integer represented by the 0 and 1 characters.           */
int get_binary_value(char *p)
{
   int value=0;                    /* Declare and initialize a temporary variable.          */

   for (; *p ; p++)                /* Keep on processing the character pointed to by        */
      value = (value<<1) + (*p - '0'); /* '*p' and incrementing '*p' until it points to a   */
   return value;                   /* null character, then return the contents of 'value'.  */
}
```

The get_binary_value()function accepts a pointer to a string of characters such as "1001", which have been read from the input file and which represent a set of input values associated with a line from one of the truth tables. The function converts this string into the numerical value that they represent; for example, "1001" = 9. The function is also used to process a single "0" or "1" character representing an output value associated with a line in a truth table, and to convert it into its numerical 0 or 1 equivalent.

The first interesting action performed by the main controlling routine is to call the function load_lookup_tables(), and it doesn't take a rocket scientist to guess what this one does:

```
/* load_lookup_tables  : Loads the 'bin_to_gray' lookup table with Gray codes, and the   */
/*                     : 'gray_to_bin' lookup table with binary codes.                    */
/*                     : Returns a dummy parameter                                        */
int load_lookup_tables()
{
    int bin, gray;                  /* Declare two temporary variables.                   */

    for (bin = 0; bin < XSIZE; bin++) { /* Increment 'bin' through a binary sequence...   */
        gray = bin ^ (bin >> 1);        /* ... 'gray' equals the Gray code equivalent of 'bin' */
        bin_to_gray[bin] = gray;        /* ... load Gray code into 'bin_to_gray' lookup table. */
        gray_to_bin[gray] = bin;        /* ... load binary value into 'gray_to_bin' lookup. */
    }

    return 0;
}
```

The load_lookup_tables() function revolves around a loop, in which the variable bin is incremented from zero to the maximum number of elements in the tables. For each value of bin, its equivalent Gray code is generated and assigned to the variable gray. Finally, the element in the bin_to_gray lookup table pointed to by bin is loaded with the value in gray, and the element in the gray_to_bin lookup table pointed to by gray is loaded with the value in bin. Note that the statement "gray = bin ^ (bin >>1)" is the software equivalent of the binary to Gray code converter illustrated in Figure D.4 in Appendix D.

Throughout the rest of this program, binary to Gray code and Gray code to binary conversions can be easily performed by accessing the appropriate value in the bin_to_gray and gray_to_bin lookup tables, respectively.

For example, assuming that an integer variable called *index* has already been assigned the equivalent of the binary value 1100_2:

```
temp = bin_to_gray[index];      {Would result in 'temp' containing the binary value 1010₂ }
temp = gray_to_bin[index];      {Would result in 'temp' containing the binary value 1000₂}
```

You can check these results using Figure D.1 in Appendix D. After everything has been declared and the lookup tables have been initialized, the program is ready to rock-and-roll. The first function of any significance is get_truth_tables(), which loads the truth tables, one at a time, from the input file:

```c
/* get_truth_tables   : Loads one truth table at a time from the input file into the working  */
/*                    : area, 'workspace'. Be very careful—this function contains minimal     */
/*                    : parsing and no error detection. As a wise man once said—"You can       */
/*                    : make it fool proof ..... but you can't make it damn-fool proof".        */
/*                    : Returns the number of inputs in the table being processed.             */
int get_truth_tables()
{
    #define XLINE 132               /* Set the max No. of characters allowed on a line.      */
    char *p;                        /* Declare a character pointer variable.                 */
    char *strtok();                 /* Declare a function (from the standard C library).     */
    char line[XLINE];               /* Reserve space to load a line from the input file.     */
    int num_in;                     /* Keeps track of the number of truth table inputs.      */
    int inputs, output;             /* Binary equivalents of the inputs and output           */
    int in_table = 0;               /* Keeps track of whether we're in a truth table         */

    while(gets(line)) {             /* Use the standard function 'gets()' to read a line.    */
        if (!(p=strtok(line," \t"))) {  /* If std function 'strtok()' says this line empty ... */
            if (in_table)           /* ...if we were already processing a table then         */
                return num_in;      /*    we've finished, so return number of inputs.        */
            else                    /* ...but if we weren't already processing a table,      */
                continue;           /*    then skip the rest of the "while" loop, go back    */
        }                           /*    to the start of the loop, and get the next line.   */

        in_table = 1;               /* Have got a non-blank line, so must be in a table.     */
        num_in = strlen(p);         /* Use std function 'strlen()' to find No. of inputs.    */
        inputs = get_binary_value(p);   /* Get the integer value represented by the inputs.  */
        p = strtok((char *)0," \t");    /* Tell 'strtok()' to get next token (the output).   */
        output = get_binary_value(p);   /* Get the integer value represented by the output.  */
        workspace[gray_to_bin[inputs]] = output; /* Load the output data into the workspace. */
    }                               /* Go back to the beginning of the loop.                 */

    return in_table ? num_in : 0;   /* Exit loop at end-of-file. If in a table then return   */
}                                   /* the number of inputs, otherwise return zero.          */
```

Functions that read-in and write-out data are generally rather boring, and this one is no exception. On the other hand, even though it is "mean and lean," get_truth_tables() permits a reasonable amount of latitude, and it's worth taking the time to note its main features and limitations:

FEATURES	LIMITATIONS
Allows any number of blank lines, including lines containing spaces and tabs, at the beginning and end of the file, and also between truth tables.	No blank lines are allowed in the middle of a truth table, but there must be at least one blank line separating one truth table from another.
Allows multiple spaces and tabs at the beginning and end of each truth table line, and also between the inputs and outputs.	All the input values must be grouped together without any spaces or tabs, but there must be at least one space or tab separating the inputs from the output.
The lines of a truth table can be specified in any order	All the lines in a truth table must be specified.
This version of the program supports truth tables with two to ten inputs.	This version of the program only supports truth tables with one output.

Actually, considering the size of this function, or its lack thereof, you couldn't really ask for too much more—what do you want for nothing, your money back? The function processes each line of a truth table and loads the output value associated with that line into the working area workspace[]. When the function detects a blank line indicating the end of the truth table, it terminates and returns the number of inputs associated with this table to the main controlling loop.

The statement of most interest is "workspace[gray_to_bin[address]] = data;", which loads the output value into the working area. This statement uses the Gray code to binary lookup table to transpose the address representing the truth table's input values into an address used by the computer to point to the working area, and to load the output value from the truth table into the element in the working area pointed to by this transposed address. The organization of the working area is the key to the way in which the rest of the program works.

The problem is that it is not easy to check for Reed-Müller implementations if the data is represented in the form of a truth table. The most efficient method (at least, that the author can discover) is to visualize and process the data in the form of a Karnaugh map. However, for a variety of reasons, it is easier for the program to handle the data as a one-dimensional array. Additionally, any solution must easily accommodate truth tables (or Karnaugh maps) of different sizes; that is, with different numbers of inputs.

The working area used by this program is a one-dimensional array of characters called workspace[]. An array of characters is used because they are only 8-bits wide, and therefore occupy less of the computer's memory than would integers, which may each require 16-bits or 32-bits depending on the compiler being used:

Figure E.2: The working area as a one-dimensional array

In fact, this program only uses the least-significant bit of each character's 8-bit field to store the contents (output values) of the truth table, and the other seven bits are therefore available for possible future enhancements. The next thing is to understand how the contents of the truth table are mapped into the working area. The easiest way to visualize this is to first view the truth table as a Karnaugh map, and then to imagine grabbing the upper-left-hand and upper-right-hand corner boxes of map and pulling them to the left and right respectively, thereby expanding the map like a concertina (Figure E.3).

Figure E.3: Mapping a truth table into the working area

The next point to note is that there are number of alternative ways of viewing the Karnaugh map associated with a particular truth table. For example, a table with four inputs may be represented as a Karnaugh map organized with four columns and four rows, two columns and eight rows, eight columns and two rows, etc. Consider a three-input truth table represented as 4 × 2 and 2 × 4 Karnaugh maps (Figure E.4).

The beauty of the technique used by this program to map the data from a truth table into the working area is that, irrespective of the size of the truth table, and irrespective of how the truth table is visualized as a Karnaugh map, the data is always mapped into the same locations in the working area (Figure E.5).

Figure E.4: Alternative Karnaugh map representations

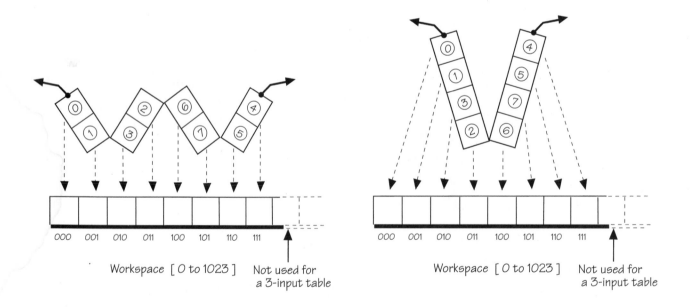

Figure E.5: Consistent mapping to the working area

Thus, the statement "workspace[gray_to_bin[address]] = data;" in the get_truth_tables() function transposes the truth table (or Karnaugh map) address, which is considered to be in the form of a Gray code, into a binary equivalent which is used to point to the working area. The output value from the truth table is loaded into the working area at this transposed address.

After all of the data from a truth table has been loaded into the working area, the function test_reed_muller() is called to determine whether this particular truth table has a Reed-Müller solution:

```
/* test_reed_muller : Checks the contents of the truth table to see if this table has a    */
/*                  : Reed-Müller solution.                                                */
/*                  : Accepts one parameter, 'num_in', the number of inputs in this table. */
/*                  : Returns 1 if this table has a Reed-Müller solution, and 0 if it doesn't. */
int test_reed_muller(int num_in)
{
    char *bottom, *top;                         /* Declare temporary character pointers.   */
    int start, end, look_for;                   /* Declare local variables.                */

    for (start = 4, end = power_two(num_in); start <= end; start <<=1) {    /* Main loop.   */
        bottom = &workspace[0];                 /* Point 'bottom' at the LHS of 'workspace'. */
        top = &workspace[start - 1];            /* Point 'top' at the RHS for this pass.     */
        look_for = *bottom++ ^ *top—;           /* Find out what we're looking for.          */
        do {                                    /* Enter a sub-loop.                         */
            if ((*bottom++ ^ *top—) != look_for) /* If the rest aren't what we want ...      */
                return 0;                       /* ... Return 0 to indicate no solution.     */
        } while (bottom < top);                 /* ... Otherwise continue with the sub-loop. */
    }

                                                /* If we get this far, then there is a solution, */
    return 1;                                   /* so return 1 to tell the world.            */
}
```

The test_reed_muller() function performs a series of actions which are more easily described pictorially. Consider the case of a two-input truth table (Figure E.6). The main loop first initializes bottom and top to point at locations 0 and 3 in the working area, respectively. The first action is to perform an exclusive-OR of the contents of the working area pointed to by bottom and top, and to store the result in the variable look_for. It doesn't actually matter whether

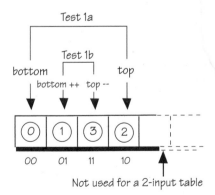

Figure E.6: Testing a two-input truth table

the result stored in look_for is a 0 or a 1; whatever it is, that's what we're looking for. The values of bottom and top are then incremented and decremented, respectively, an exclusive-OR is performed on the contents of the working area that they are now pointing to, and the result is compared to the value stored in look_for.

If the result of this exclusive-OR is not the same as the value stored in look_for, then there is no Reed-Müller solution, and test_reed_muller() returns a 0 and terminates. But, if the result is the same as the value stored in look_for, then the function continues onward. In fact, in the case of a two-input truth table, all of the tests have been performed, so the function would return a 1 indicating that there is a Reed-Müller solution. Now consider the case of a three-input truth table:

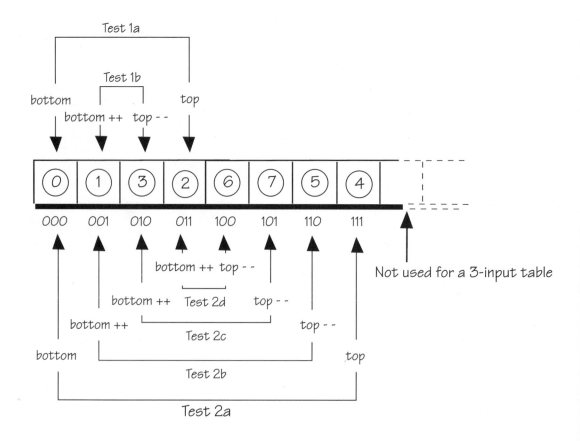

Figure E.7: Testing a three-input truth table

The first series of tests are exactly the same for a two-input truth table. Assuming that the table passes these tests, the main loop reinitializes bottom and top to point at locations 0 and 7 in the working area respectively. Once again, the first action in this iteration of the loop is to perform an exclusive-OR of the contents of the working area pointed to by bottom and top, and to store the result in look_for. And once again, it doesn't actually matter whether the result stored in look_for is a 0 or a 1; whatever it is, that's what we're looking for this time around.

The values of bottom and top are then incremented and decremented respectively, an exclusive-OR is performed on the contents of the working area that they are now pointing to, and the result is compared to the value stored in look_for. This continues until it is determined that there is no Reed-Müller solution, or until bottom and top meet in the middle, in which case there is a solution.

Similarly, in the case of a four-input truth table, the first series of tests would be the same as for a two-input table, and the second series of tests would be the same as for a three-input table. Assuming that the table passes these tests, the main loop would commence a third series of tests by reinitializing bottom and top to point at locations 0 and 15 in the working area respectively. As usual, the first action in this third iteration of the loop would be to perform an exclusive-OR of the contents of the working area pointed to by bottom and top, and to store the result in look_for. The process would then continue as before: incrementing bottom, decrementing top, and testing their contents against the value in look_for until it is determined that there is no Reed-Müller solution, or until bottom and top meet in the middle, in which case there is a solution.

The problem is that, although these tests are simple, there can be a lot of them. On the positive side, if there is no Reed-Müller solution then this function usually finds out fairly quickly. But, if there is a solution, then it can take a lot of tests to prove it. In fact, an n-input truth table will require $(2^n - 2)$ tests if it has a solution. This is not ideal, but there does not appear to be a better way ... unless you can find one (more below). Assuming that there is a Reed-Müller solution, then the extract_reed_muller() function is used to determine the significant inputs:

```
/* extract_reed_muller  : Extracts the significant inputs for the Reed-Müller solution.   */
/*                       : Accepts one parameter, 'num_in', the No. of inputs in the table. */
/*                       : Returns a bitmask, 'mask', representing the significant inputs.  */
int extract_reed_muller(int num_in)
{
    int i, x, y, mask, index, look_for;      /* Declare all the local variables.          */

    mask = power_two(num_in) - 1;            /* Set 'mask' to contain 1s for each input.  */
    look_for = workspace[0];                 /* Set 'look_for' to the value in [0,0] of the map. */
    x = xy_table[num_in].x;                  /* Get the most appropriate number of columns. */
    y = xy_table[num_in].y;                  /* Get the most appropriate number of rows.   */

    /* The following routines view the working area as a Karnaugh map. When I wrote        */
    /* them, only God and myself knew how they worked .... now God only knows!             */

    for (i = 1; i < x; i++) {                /* This loop commences at box [0,0] in the map */
      if (!(i & 1))                          /* (the upper-left corner) and searches "East". */
        index = i * y;                       /* For every box on its journey it, must generate */
      else                                   /* a value for 'index', which acts as a pointer */
        index = ((i + 1) * y) - 1;           /* into the working area.                     */

      if (workspace[index] == look_for)      /* If the box contains the same as 'look_for' ... */
        mask &= ~(bin_to_gray[index]);       /* ... Remove non-significant input from 'mask'. */
      else                                   /* ... Otherwise break out of this loop and move */
        break;                               /* onto the next one.                         */
    }

    if (i < (x >> 1)) {                      /* Before beginning this loop, a test is performed */
      for (i = (x - 1); i > 0; i—) {         /* to see if it's actually necessary. If the last */
        if (!(i & 1))                        /* loop searched half-way, or all the way across */
          index = i * y;                     /* the map, then there's no point in performing */
        else                                 /* the tests in this loop.                    */
          index = ((i + 1) * y) - 1;

                                             /* Otherwise, this loop searches "West"      */
        if (workspace[index] == look_for)    /* from box [0,0] in the map. Because of      */
          mask &= ~(bin_to_gray[index]);     /* the wrap-around effect associated with     */
        else                                 /* Karnaugh maps, this means that the         */
          break;                             /* search actually commences in the           */
      }                                      /* upper-right corner of the map, and         */
    }                                        /* then heads back towards box [0,0].         */

    for (i = 1; i < y; i++) {                /* This loop commences at box [0,0], and      */
      if (workspace[i] == look_for)          /* and searches "South"; that is, towards the */
        mask &= ~(bin_to_gray[i]);           /* bottom-left corner.                        */
      else
        break;
    }
```

```
if (i < (y >> 1)) {              /* Tests to see if this loop is actually necessary.  */
    for (i = (y - 1); i > 0; i—) {  /* If it is, then search "North" from box [0,0].   */
        if (workspace[i] == look_for)  /* Because of the Karnaugh map wrap-around       */
            mask &= ~(bin_to_gray[i]);  /* effect, the search actually commences in the  */
        else                          /* lower-left corner of the map, and then heads    */
            break;                    /* back towards box [0,0].                         */
    }
}

    return mask;                     /* Finally, return the significant inputs in the   */
}                                    /* form of the bitmask, 'mask'.                    */
```

The extract_reed_muller() function is the hardest to follow, but it is certainly a good exercise in reverse-polish-logic-lateral-thinking which will keep your brain in peak mental fitness. The easiest way to visualize the East-West-South-North searches is shown below. Consider the Karnaugh map associated with a four-input truth table:

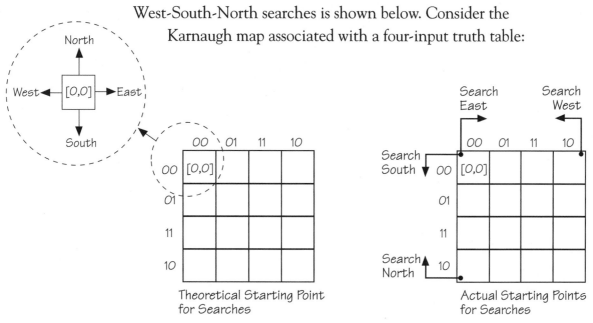

Figure E.8: Visualizing the searches to locate the significant inputs

In fact, the extract_reed_muller() function is reasonably efficient. One of the rules for a Reed-Müller solution is that there must be the same number of 1s stored in the Karnaugh map boxes as there are 0s. This means that the largest possible block of 1s or 0s can only occupy half of the boxes, which in turn means that this function actually performs relatively few tests. In fact, for a four-input truth table, this function will only ever performs a maximum of four tests (excluding the tests to see whether further tests are necessary). Now it only remains to output the equation for the Reed-Müller implementation:

```
/* output_reed_muller   : Output the equation for the Reed-Muller implementation.   */
/*                       : Accepts three parameters: 'mask' which contains the significant */
/*                       : inputs, 'num_tt' which contains the No. of the truth table in the */
/*                       : input file, and 'num_in' which contains the No. of table inputs.  */
/*                       : Returns a dummy value.                                    */
int output_reed_muller(int mask, int num_tt, int num_in)
{
  int i;                          /* Declare a temporary integer variable.      */
  char c;                         /* Declare a temporary character variable.    */

  if (!workspace[O])                                 /* If box [0,0] holds    */
    printf("Truth table %i: Reed-Muller solution is y = (",num_tt);  /* a zero, then only     */
  else                                               /* need XORs, else       */
    printf("Truth table %i: Reed-Muller solution is y = !(",num_tt); /* add an inversion.     */

  for (i = (power_two(num_in) >> 1), c = 'a'; !(mask & i); i >>= 1, c++);
  printf("%c ",c);
  for (i >>= 1, c++; i > O; i >>= 1, c++)    /* Explaining this rather cunning routine    */
    if (mask & i)                            /* is left as an exercise for the reader ... */
      printf("^ %c ",c);                     /* ... Ha!                                   */
  printf(")\n");

  return O;                       /* Return a dummy value.                     */
}
```

How to Become Famous

There are a number of ways in which the program described above could be modified or extended. For example, the input function, get_truth_tables(), could be easily modified to accept truth tables with multiple outputs. Alternatively, it could be augmented to accept the textual equivalent of Karnaugh maps. A more interesting exercise would be to modify the program to accept don't care values (represented using question marks, "?") on the inputs (Figure E.9).

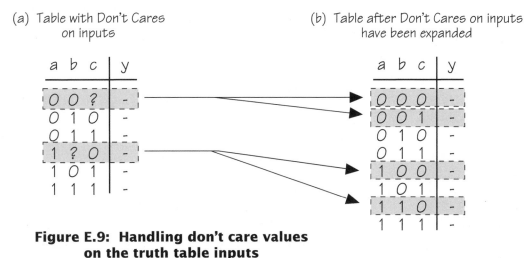

(a) Table with Don't Cares on inputs

(b) Table after Don't Cares on inputs have been expanded

Figure E.9: Handling don't care values on the truth table inputs

This modification would be relatively simple because it would only affect the input function. Note that handling a single don't care value on a line of inputs is trivial, but handling multiple don't cares is a little trickier. For those who enjoy a real challenge, modifying the program to cope with don't cares on the outputs[5] should offer one! This is where the spare bits that are available in the working area may start to earn their keep.

Another interesting area to explore would be the test_reed_muller() function. As was noted during the discussions on this function, an n-input truth table will require $(2^n - 2)$ tests (assuming that the table has a Reed-Müller solution). It may be that you can discover a more efficient technique to determine whether the table has a solution.[6]

During the course of this discussion, the question may have arisen as to why the maximum number of truth table inputs is set as high as ten. After all, this would equate to a truth table with 1,024 lines, and there is little chance that anyone would take the time to type these in (and even less chance of their doing so without errors). However, one way in which

---

[5] Note that the concepts of don't cares on inputs and outputs were introduced in more detail in Chapter 10.

[6] Beware when you alter this function, there are many traps for the unwary. A useful test is to apply your routine to the four-input truth table equivalent of a 4 × 4 Karnaugh map which contains only four 1s, one in each corner. If your routine decides that this table has a Reed-Müller solution, then it's wrong!

the program could be modified would be for it to read in simple Boolean equations. For example, the input function could be modified to process the following text file:

```
INPUTS          = a, b, c, d;
OUTPUTS         = x, y;
INTERNAL_NODES  = j, k;


j =    (a & b);
k = a | (c ^ d);
x = j & (c | d);
y =    !(k & b);


END
```

The format shown here is fairly generic, but you could invent your own if you wish. The point is that, if you start processing equations in this form, you can easily reach the ten-input limit.

In fact, extending the program to accept Boolean equations is a large stride into the arena of *Logic Synthesis*, in which programs are used to optimize the logic used to implement a design. Many commercially available Logic Synthesis utilities use a method of representing the data internally known as binary decision diagrams (BDDs), in which equations are stored in decision trees.

For example, consider the binary decision diagram for the simple expression "y = (a & b) | c" (Figure E.10). The circles representing the inputs are known as nodes, and the 0 and 1 values associated with the lines linking two nodes indicate the possible state of the upper node. The 0 and 1 values at the bottom of the diagram indicate the results of the equation for each combination of inputs. The initial node is known as the root, the links between nodes may be referred to as branches, and the lowest links and their values are referred to as leafs.[7]

The binary decision diagram form of representation finds many uses. For example, as an alternative to actually evaluating a complex expression,

---

[7] This terminology makes sense if the diagram is visualized as an inverted tree.

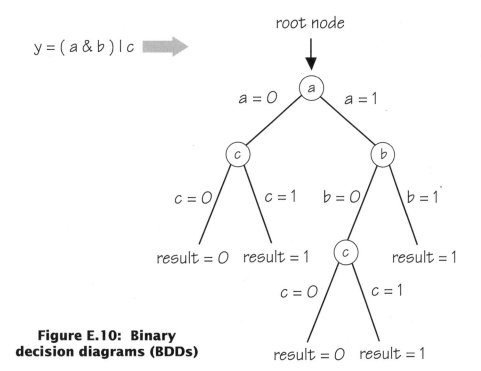

$y = (a \& b) \mid c$

root node

Figure E.10: Binary decision diagrams (BDDs)

it is possible to quickly determine the output from the expression by marching through the diagram using the values on the inputs as signposts. This is of interest because it may be possible to use these diagrams as inputs to Max's Algorithm. As the size of an expression increases, it becomes increasingly unlikely that the entire expression will have a Reed-Müller solution. However, it is more than likely that some portions of the expression may have such solutions. Under the direction of a controlling function, it would be possible to examine the decision diagram for Reed-Müller solutions working from the leafs upwards.

You may feel free to use this program in your own projects. However, it would be nice if you could reference Max's Algorithm and its source, *Bebop to the Boolean Boogie*. Additionally, if you do adopt any of the suggestions above, or make modifications of your own, or use the algorithm for a real application, please contact the author (care of the publisher) and spread a little joy in the world. The most interesting derivations may well find themselves featured in the next edition of *Bebop*, thereby making you famous beyond your wildest dreams (and your mother very proud).

Linear Feedback Shift Registers (LFSRs)

Linear feedback shift registers (LFSRs) are simple to construct and are useful for a wide variety of applications, but are often sadly neglected by designers. An LFSR is formed from a simple shift register with feedback from two or more points, or *taps*, in the register chain:

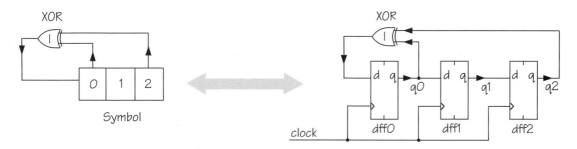

Figure F.1: LFSR with XOR feedback path

The taps in this example are at bit 0 and bit 2, and can be referenced as [0,2]. All of the register elements share a common clock which is omitted from the symbol for reasons of clarity. The data input to the LFSR is generated by XOR-ing or XNOR-ing the tap bits; the remaining bits function as a standard shift register. The sequence of values generated by an LFSR is determined by its feedback function (XOR versus XNOR) and tap selection (Figure F.2).

Both LFSRs start with the same initial value but, due to the different taps, their sequences rapidly diverge as clock pulses are applied. After the second clock, the LFSR with taps at [1,2] enters a loop comprising only two values. By comparison, the LFSR with taps at [0,2] is said to be of *maximal-length* because it sequences through every possible value (excluding all the bits being 0) before returning to its initial value. A binary field with *n* bits

Clock Edge	q0	q1	q2	
—	0	0	1	← Initial Value
1	1	0	0	
2	0	1	0	
3	1	0	1	
4	0	1	0	
5	1	0	1	
6	0	1	0	
7	1	0	1	
8	0	1	0	

Clock Edge	q0	q1	q2	
—	0	0	1	← Initial Value
1	1	0	0	
2	1	1	0	
3	1	1	1	
4	0	1	1	
5	1	0	1	
6	0	1	0	
7	0	0	1	
8	1	0	0	

Figure F.2: Comparison of alternative tap selections

Clock Edge	q0	q1	q2	
—	0	0	1	← Initial Value
1	1	0	0	
2	1	1	0	
3	1	1	1	
4	0	1	1	
5	1	0	1	
6	0	1	0	
7	0	0	1	
8	1	0	0	

Clock Edge	q0	q1	q2	
—	0	0	1	← Initial Value
1	0	0	0	
2	1	0	0	
3	0	1	0	
4	1	0	1	
5	1	1	0	
6	0	1	1	
7	0	0	1	
8	0	0	0	

Figure F.3: Comparison of XOR and XNOR feedback paths

can assume 2^n unique values, but a maximal-length LFSR with n register bits will only sequence through $(2^n - 1)$ values. LFSRs with XOR feedback paths will not sequence through the value where all the bits are 0, while their XNOR equivalents will not sequence through the value where all the bits are 1 (Figure F.3).

The author created a simple program to determine the taps for maximal-length LFSRs with 2 to 32 bits; the * annotations indicate sequences whose length is a prime number (Figure F.4).

The taps are identical for both XOR-based and XNOR-based LFSRs. Alternative tap combinations may also yield maximum-length LFSRs, although the resulting sequences will vary. For example, in the case of a 10-bit LFSR, there are 2 × 2-tap combinations that result in a maximal-length sequence: [2,9] and [6,9]. There are also 20 × 4-tap combinations, 28 × 6-tap combinations, and 10 × 8-tap combinations that satisfy the maximal-length criteria.

# of Bits	Length of Loop		Taps
2	3	*	[0,1]
3	7	*	[0,2]
4	15		[0,3]
5	31	*	[1,4]
6	63		[0,5]
7	127	*	[0,6]
8	255		[1,2,3,7]
9	511		[3,8]
10	1,023		[2,9]
11	2,047		[1,10]
12	4,095		[0,3,5,11]
13	8,191	*	[0,2,3,12]
14	16,383		[0,2,4,13]
15	32,767		[0,14]
16	65,535		[1,2,4,15]
17	131,071	*	[2,16]
18	262,143		[6,17]
19	524,287	*	[0,1,4,18]
20	1,048,575		[2,19]
21	2,097,151		[1,20]
22	4,194,303		[0,21]
23	8,388,607		[4,22]
24	16,777,215		[0,2,3,23]
25	33,554,431		[2,24]
26	67,108,863		[0,1,5,25]
27	134,217,727		[0,1,4,26]
28	268,435,455		[2,27]
29	536,870,911		[1,28]
30	1,073,741,823		[0,3,5,29]
31	2,147,483,647	*	[2,30]
32	4,294,967,295		[1,5,6,31]

Figure F.4: Taps for maximal length LFSRs with 2 to 32 bits

Seeding an LFSR

One quirk with an XOR-based LFSR is that, if it happens to find itself in the all 0s value, it will happily continue to shift all 0s indefinitely; similarly for an XNOR-based LFSR and the all 1s value. This is of particular concern when power is first applied to the circuit. Each register bit can randomly initialize containing either a logic 0 or a logic 1, and the LFSR can therefore wake up containing its "forbidden" value. For this reason, it is necessary to initialize an LFSR with a *seed* value.

One method for loading a seed value is to use registers with *reset* or *set* inputs. A single control signal can be connected to the *resets* on some of the registers and the *sets* on others. When this control signal is placed in its active state, the LFSR is loaded with a hard-wired seed value. However, in certain applications it is desirable to be able to vary the seed value. One technique for achieving this is to include a multiplexer at the input to the LFSR (Figure F.5).

Figure F.5: Loading alternative seed values

When the *data* input is selected, the device functions as a standard shift register and any desired seed value may be loaded. After loading the seed value, the feedback path is selected and the device returns to its LFSR mode of operation.

FIFO Applications

The fact that an LFSR generates an unusual sequence of values is irrelevant in many applications. Consider a 4-bit × 16-word first-in first-out (FIFO) memory device (Figure F.6).

A brief summary of the operation of the FIFO follows. The write and read pointers are essentially 4-bit registers whose outputs are processed by 4:16 decoders to select one of the sixteen words in the memory array. The *reset* input is used to initialize the device, primarily by clearing the write and read pointers such that they both point to the same memory word. The initialization also causes the *empty* output to be placed in its active state and the *full* output to be placed in its inactive state.

Figure F.6: First-in first-out (FIFO) memory

The write and read pointers chase each other around the memory array in an endless loop. An active edge on the *write* input causes any data on the *data-in* bus to be written into the word pointed to by the write pointer; the *empty* output is placed in its inactive state and the write pointer is incremented to point to the next empty word. Data can be written into the FIFO until all the words in the array contain data. When the write pointer catches up to the read pointer, the *full* output is placed in its active state and no more data can be written into the device.

An active edge on the *read* input causes the data in the word pointed to by the read pointer to be copied into the output register; the *full* output is placed in its inactive state and the read pointer is incremented to point to the next word containing data.[1] Data can be read out of the FIFO until the array is empty. When the read pointer catches up to the write pointer, the

---

[1] These discussions assume *write-and-increment* and *read-and-increment* techniques. However, some FIFOs use an *increment-and-write* and *increment-and-read* approach.

empty output is placed in its active state and no more data can be read out of the device.

The write and read pointers for a 16-word FIFO are often implemented using 4-bit binary counters.[2] However, a moment's reflection reveals that there is no intrinsic advantage to a binary sequence for this particular application; the sequence generated by a 4-bit LFSR would serve equally well. In fact, the two functions operate in a very similar manner as is illustrated by their block diagrams:

Figure F.7: Binary counter versus LFSR

However, the combinational logic for the 4-bit LFSR consists of a single two-input XOR gate, while the combinational logic for the 4-bit binary counter requires a number of AND and OR gates. This means that the LFSR requires fewer tracks and is significantly more efficient in terms of silicon real estate. Additionally, the LFSR's feedback only passes through a single level of logic, while the binary counter's feedback passes through multiple levels of logic. This means that the new data value is available sooner for the LFSR which can therefore be clocked at a higher frequency. These differentiations become even more pronounced for FIFOs with more words requiring pointers with more bits. Thus, LFSR's are the obvious choice for the discerning designer of FIFOs.

---

[2] The implementation of a 4-bit binary counter was discussed in Chapter 11.

Modifying LFSRs to Sequence 2^n Values

The sole downside to the 4-bit LFSRs in the FIFO scenario above is that they will only sequence through 15 values ($2^4 - 1$), as compared to the binary counter's sequence of 16 values (2^4). Designers may not regard this to be a problem, especially in the case of larger FIFOs. However, if it is required for an LFSR to sequence through every possible value, there is a simple solution (Figure F.8).

For the value where all the bits are 0 to appear, the preceding value must have comprised a logic 1 in the MSB[3] and logic 0s in the remaining bit positions. In an unmodified LFSR, the next clock would result in a logic 1 in the LSB and logic 0s in the remaining bit positions. However, in the modified LFSR, the output from the NOR is a logic 0 for every case except the value preceding the one where all the bits are 0 and the value where all the bits are 0. These two values force the NOR's output to a logic 1 which inverts the usual output from the XOR. This in turn causes the sequence to first enter the all 0s value and then resume its normal course. In the case of LFSRs with XNOR feedback paths, the NOR can be replaced with an AND which causes the sequence to cycle through the value where all the bits are 1.

Clock Edge	q0	q1	q2	q3	
—	0	0	0	1	
1	0	0	0	0	← All 0's
2	1	0	0	0	
3	1	1	0	0	
4	1	1	1	0	
5	1	1	1	1	
6	0	1	1	1	
7	1	0	1	1	
8	0	1	0	1	2^4=16 Values
9	1	0	1	0	
10	1	1	0	1	
11	0	1	1	0	
12	0	0	1	1	
13	1	0	0	1	
14	0	1	0	0	
15	0	0	1	0	
16	0	0	0	1	

Figure F.8: LFSR modified to sequence 2^n values

---

[3] As is often the case with shift registers, the MSB in these examples is taken to be on the right-hand side of the register and the LSB is taken to be on the left-hand side.

Accessing the Previous Value

In some applications it is required to make use of a register's previous value. For example, in certain FIFO implementations, the full condition is detected when the write pointer is pointing to the location preceding the location pointed to by the read pointer.[4] This implies that a comparator must be used to compare the current value in the write pointer with the previous value in the read pointer. Similarly, the *empty* condition may be detected when the read pointer is pointing to the location preceding the location pointed to by the write pointer. This implies that a second com-parator must be used to compare the current value in the read pointer with the previous value in the write pointer.

In the case of binary counters, there are two techniques by which the previous value in the sequence may be accessed. The first technique requires the provision of an additional set of shadow registers; every time the counter is incremented, the current contents are first copied into the shadow registers. Alternatively, a block of combinational logic can be used to decode the previous value from the current value. Unfortunately, both of these techniques involve a substantial overhead in terms of additional logic. By comparison, LFSRs inherently remember their previous value; all that is required is the addition of a single register bit appended to the MSB:

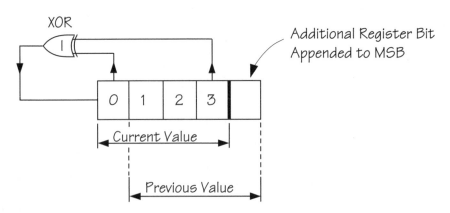

Figure F.9: Accessing an LFSR's previous value

---

[4] Try saying that quickly!

Encryption and Decryption Applications

The unusual sequence of values generated by an LFSR can be gainfully employed in the encryption (scrambling) and decryption (unscrambling) of data. A stream of data bits can be encrypted by XOR-ing them with the output from an LFSR:

Figure F.10: Data encryption using an LFSR

The stream of encrypted data bits seen by a receiver can be decrypted by XOR-ing them with the output of an identical LFSR.

Cyclic Redundancy Check Applications

A traditional application for LFSRs is in cyclic redundancy check (CRC) calculations which can be used to detect errors in data communications. The stream of data bits being transmitted is used to modify the values fed back into an LFSR (Figure F.11).

The final CRC value stored in the LFSR is known as a *checksum*, and is dependent on every bit in the data stream. After the data bits have been transmitted, the transmitter sends its

Figure F.11: Cyclic redundancy check (CRC) calculation

checksum value to the receiver. The receiver contains an identical CRC calculator and generates its own checksum value from the incoming data. The receiver compares its internally generated checksum value with the checksum sent by the transmitter to determine whether any corruption occurred during the course of the transmission. This form of error detection is very efficient in terms of the small number bits that have to be transmitted in addition to the data.

In the real world, a 4-bit CRC calculator would not be considered to provide sufficient confidence in the integrity of the transmitted data. Because a 4-bit LFSR can only represent 16 unique values, there is a significant probability that multiple errors in the data stream could result in the two checksum values being identical. As the number of bits in a CRC calculator increases, the probability that multiple errors will cause identical checksum values approaches zero. For this reason, CRC calculators typically use a minimum of 16-bits providing 65,536 unique values.

There are a variety of standard communications protocols, each of which specifies the number of bits employed in their CRC calculations and the taps to be used. The taps are selected such that an error in a single data bit will cause the maximum possible disruption to the resulting checksum value. Thus, in addition to maximal-length, these LFSRs may also be qualified as *maximal-displacement*.

In addition to checking data integrity in communications systems, CRCs find a wide variety of other uses; for example, the detection of computer viruses. There are many different types of computer viruses but, for the purposes of this discussion, a computer virus may be defined as a self-replicating program released into a computer system for a number of purposes. These purposes range from the simply mischievous, such as displaying humorous or annoying messages, to the downright nefarious, such as corrupting data or destroying (or subverting) the operating system.

One mechanism by which a computer virus may both hide and propagate itself is to attach itself to an existing program. Whenever that program is executed it first triggers the virus to replicate itself, yet a cursory check of the system only shows expected files to be present. To combat this form of attack, a unique checksum can be generated for each program on the system, where the value of each checksum is based on the binary instructions forming the program with which it is associated. At some later date, an anti-virus program can be used to recalculate the checksum values for each program, and to compare them to the original values. A difference in the two values for a program may indicate that a virus has attached itself to that program.[5]

Data Compression Applications

The CRC calculators discussed above can also be used in a data compression role. One such application is found in the circuit board test strategy known as *functional test*. The board is plugged into a functional tester by means of its edge connector. The tester applies a pattern of signals to the board's inputs, allows sufficient time for any effects to propagate around the board, and then compares the actual values seen on the outputs with a set of expected values stored in the system. This process is repeated for a series of input patterns which may number in the thousands:

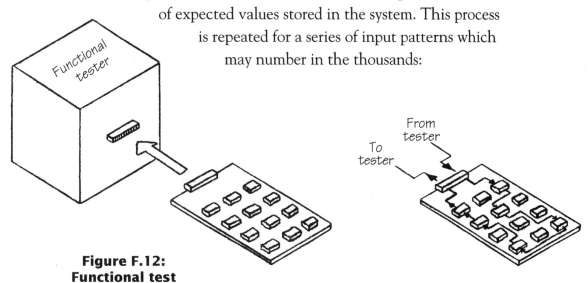

Figure F.12:
Functional test

The illustration above is simplified for reasons of clarity. In practice, the edge connector may contain hundreds of pins while the board may contain thousands of components and tracks. If the board fails the preliminary tests, a more sophisticated form of analysis known as *guided probe* may be employed to identify the cause of the failure (Figure F.13).

---

[5] Unfortunately, the creators of computer viruses are becoming more sophisticated, and some viruses are armed with the ability to perform their own CRC calculations. When a virus of this type attaches itself to a program, it can pad itself with dummy binary values which will cause an anti-virus program to return an identical checksum value to the original.

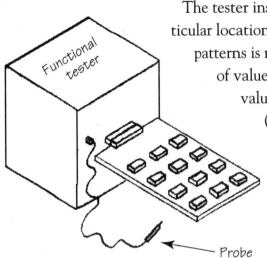

Figure F.13: Guided probe analysis

The tester instructs the operator to place the probe at a particular location on the board, then the entire sequence of test patterns is rerun. The tester compares the actual sequence of values seen by the probe with a sequence of expected values which are stored in the system. This process (placing the probe and running the tests) is repeated until the tester has isolated the faulty component or track.

A major consideration when supporting a guided probe strategy is the amount of expected data that must be stored. Consider a test sequence comprising 10,000 patterns driving a board containing 10,000 tracks. If the data were not compressed, the system would have to store 10,000 bits of expected data per track which amounts to 100,000,000 bits of data for the board. Additionally, for each application of the guided probe, the tester would have to compare the 10,000 data bits observed by the probe with the 10,000 bits of expected data stored in the system. Thus, using data in an uncompressed form is an expensive option in terms of storage and processing requirements.

One solution to these problems is to employ LFSR based CRC calculators. The sequence of expected values for each track can be passed through a 16-bit CRC calculator implemented in software. Similarly, the sequence of actual values seen by the guided probe can be passed through an identical CRC calculator implemented in hardware. In this case, the calculated checksum values are also known as *signatures*, and a guided probe process based on this technique is known as *signature analysis*. Irrespective of the number of test patterns used, the system only has to store two bytes of data for each track. Additionally, for each application of the guided probe, the tester only has to compare the two bytes of data gathered by the probe with two bytes of expected data stored in the system. Thus, compressing the data results in storage requirements that are orders of magnitude smaller and comparison times that are orders of magnitude faster than the uncompressed data approach.

Built-In Self-Test Applications

One test strategy which may be employed in complex integrated circuits is that of built-in self-test (BIST). Devices using BIST contain special test generation and result gathering circuits:

Figure F.14: Built-in self-test (BIST)

A multiplexer is used to select between the standard inputs and those from the test generator. A second multiplexer selects between the standard outputs and those from the results gatherer. Both the test generator and results gatherer can be implemented using LFSRs (Figure F.15[6]).

The LFSR forming the test generator is used to create a sequence of test patterns, while the LFSR forming the results gatherer is used to capture the results. The results gathering LFSR contains modifications allowing it to accept parallel data. Note that the two LFSRs are not obliged to contain the same number of bits.

---

[6] The standard inputs and outputs along with their multiplexers have been omitted from this illustration for reasons of clarity.

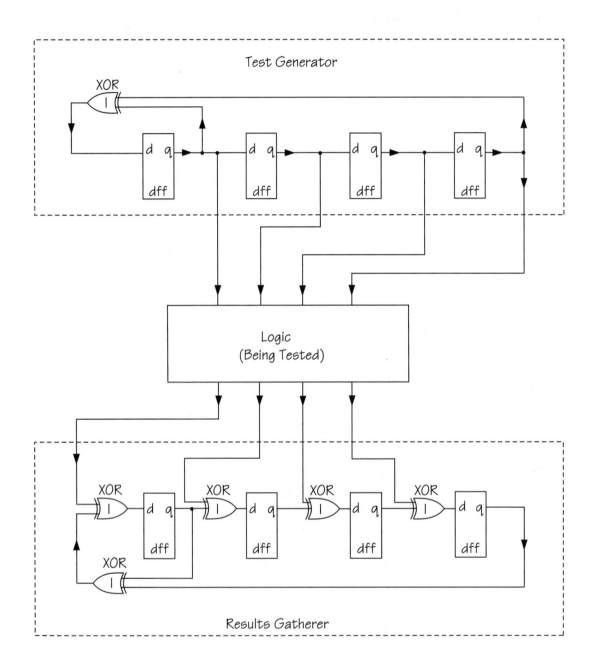

Figure F.15: BIST implemented using LFSRs

Pseudo-Random Number Applications

Many computer programs rely on an element of randomness. Computer games such as "Space Invaders" employ random events to increase the player's enjoyment. Graphics programs may exploit random numbers to generate intricate patterns. All forms of computer simulation may utilize random numbers to more accurately represent the real world. For example, digital simulations may benefit from the portrayal of random stimulus such as external interrupts. Random stimulus can result in more realistic design verification, which can uncover problems that may not be revealed by more structured tests.

Random number generators can be constructed in both hardware and software. The majority of these generators are not truly random, but they give the appearance of being random and are therefore said to be *pseudo-random*. In fact, pseudo-random numbers have an advantage over truly random numbers because the majority of computer applications typically require repeatability. For example, a designer repeating a digital simulation would expect to receive identical answers to those from the previous run. However, designers also need the ability to modify the seed value of the pseudo-random number generator so as to spawn different sequences of values as required.

There are a variety of methods available for generating pseudo-random numbers. A popular *cheap-and-cheerful* technique uses the remainder from a division operation as the next value in a pseudo-random sequence. For example, a C function which returns a pseudo-random number in the range 0 to 32767 could be written as:[7]

```
int rand ()
{
    next = next * 1103515245 + 12345;
    return (unsigned int) (next/65535) % 32768;
}
```

---

[7] *The C Programming Language* (Second Edition) by Brian W. Kernighan and Dennis M. Ritchie.

The variable *next* would be declared as global and initialized with some seed value. Also, an additional function would be provided to allow the programmer to load *next* with a new seed value if required. Every time the function is accessed, it returns a new pseudo-random value. Unfortunately, pseudo-random implementations based on division do not always produce a "white spectrum". The series of generated values may be composed of sub-series in which case the results will not be of maximal-length. By comparison, the sequence of values generated by a software implementation of a maximal-length LFSR provides an ideal pseudo-random source, but is somewhat more expensive in terms of processing requirements.

A simple technique for checking the "randomness" of a pseudo-random number generator is as follows. Assume the existence of a function called rand() which returns a pseudo-random number as a 16-bit value. The most-significant and least-significant bytes may be considered to represent the y-axis and x-axis of a graph respectively. To test the randomness of the function, a simple C program segment may be created to plot a reasonable sample of random numbers:

```
int x_axis, y_axis, i, my_number;           /* Declare the variables.                */
for (i = 1; i <= 1000; i++) {                /* Perform a loop 1,000 times.           */
    my_number = rand();                      /* Get a pseudo-random number.           */
    x_axis = my_number & 0x00FF;             /* Extract the least-significant byte.    */
    y_axis = (my_number & 0xFF00) >> 8;      /* Extract the most-significant byte.     */
    plot(x_axis, y_axis);                    /* Plot the value.                       */
}
```

This example assumes the existence of a function plot() that will cause the point represented by the variables x_axis and y_axis to be displayed on a computer screen. Note that the masking operation used when extracting the most-significant byte is included for reasons of clarity, but is in fact redundant.

If the resulting points are uniformly distributed across the plotting area, then the pseudo-random generator is acceptable. But, if the points are grouped around the diagonal axis or form clusters, then the generator is less than optimal (Figure F.16).

(a) Good

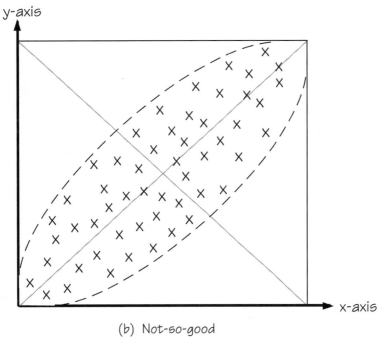

(b) Not-so-good

Figure F.16: Checking for "randomness"

Pass-Transistor Logic

The term pass-transistor logic refers to techniques for connecting MOS transistors such that data signals pass between their source and drain terminals. These techniques minimize the number of transistors required to implement a function, but they are not recommended for the novice or the unwary because strange and unexpected effects can ensue. Pass-transistor logic is typically employed by designers of full-custom integrated circuits or ASIC cell libraries. The following examples introduce the concepts of pass-transistor logic, but they are not intended to indicate recommended design practice.

Figure G.1: Pass-transistor implementation of an AND gate

In the case of the AND, the resistance of R_1 is assumed to be sufficiently high that its effect on output y is that of a very weak logic 0. When input b is set to logic 0, the NMOS transistor Tr_1 is turned OFF and y is "pulled down" to logic 0 by resistor R_1. When input b is set to logic 1, Tr_1 is turned ON and output y is connected to input a. In this case, a logic 0 on input a leaves y at logic 0, but a logic 1 on a will overdrive the effect of resistor R_1 and force y to a logic 1.

Figure G.2: Pass-transistor implementation of an OR gate

In the case of the OR, a logic 1 applied to input *b* turns the PMOS transistor Tr₁ OFF, and output y is "pulled up" to logic 1 by resistor R₁. When input *b* is set to logic 0, Tr₁ is turned ON and output y is connected to input *a*. In this case, a logic 1 on input *a* leaves y at logic 1, but a logic 0 on *a* will overdrive the effect of resistor R₁ and force y to a logic 0.

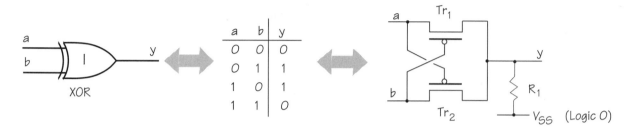

Figure G.3: Pass-transistor implementation of an XOR gate

In the case of the XOR, logic 1s applied to inputs *a* and *b* turn Tr₁ and Tr₂ OFF respectively, leaving output y to be "pulled down" to logic 0 by resistor R₁. When input *b* is set to logic 0, Tr₁ is turned ON and output y is connected to input *a*. In this case, a logic 0 on input *a* leaves y at logic 0, but a logic 1 on *a* will overdrive the effect of resistor R₁ and force y to a logic 1. Similarly, when input *a* is set to logic 0, Tr₂ is turned ON and output y is connected to input *b*. In this case, a logic 0 on input *b* leaves y at logic 0, but a logic 1 on *b* will overdrive the effect of resistor R₁ and force y to a logic 1.

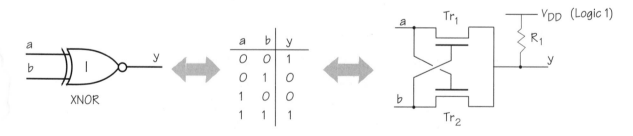

Figure G.4: Pass-transistor implementation of an XNOR gate

The case of the XNOR is simply the inverse of that for the XOR. Logic 0s applied to inputs *a* and *b* turn transistors Tr₁ and Tr₂ OFF, leaving output y to be "pulled up" to logic 1 by resistor R₁. When input *b* is set to logic 1, Tr₁ is turned ON and output y is connected to input *a*. In this case, a logic 1 on input *a* leaves *y* at logic 1, but a logic 0 on *a* will overdrive the effect of resistor R_1 and force *y* to a logic 0. Similarly, when input *a* is set to logic 1, Tr_2 is turned ON and output *y* is connected to input *b*. In this case, a logic 1 on input *b* leaves *y* at logic 1, but a logic 0 on *b* will overdrive the effect of resistor R_1 and force *y* to a logic 0.

Figure G.5: Pass-transistor implementation of a D-type latch

Before examining this particular implementation of a D-type latch with an active-high enable, it is appropriate to briefly review our definition of these functions. First, when the *enable* input is in its active state, the value on the *data* input is passed through the device to appear on the *q* and *qb* outputs.[1] Second, if the *data* input changes while *enable* remains active, the outputs will respond to reflect the new value. Third, when *enable* is driven to its inactive state, the outputs remember their previous values and no longer respond to any changes on the *data* input.

Now onto the pass-transistor implementation.[2] When *enable* is in its active state, the output from *g2* turns transistor Tr_1 ON, and the output from *g3* turns transistor Tr_2 OFF. Thus, the value on the *data* input passes through *g1*, Tr_1, and *g6* to appear on the *q* output, and through *g1*, Tr_1, *g4*, and *g7* to

---

[1] As usual, the *data* appears in inverted form on the complementary *qb* output.

[2] Note the use of assertion-level logic symbols in the circuit diagram. These symbols were introduced in Appendix A.

appear at the ~q output.[3] Any changes on *data* will be reflected on q and ~q after the delays associated with the gates and transistors have been satisfied. When *enable* is driven to its inactive state, the output from $g2$ turns Tr_1 OFF and the output from $g3$ turns Tr_2 ON. Disabling Tr_1 blocks the path from the data input, while enabling Tr_2 completes the self-sustaining loop formed by $g4$ and $g5$. It is this loop that acts as the memory for the device.

Now compare this implementation with its standard counterpart introduced in Chapter 11. The standard implementation required one NOT, two ANDs, and two NORs totaling twenty-two transistors. By comparison, the pass-transistor implementation requires two discrete transistors and seven NOTs totaling only sixteen transistors. In fact, the heart of the pass transistor implementation consisting of Tr_1, Tr_2, $g3$, $g4$, and $g5$ requires only eight transistors. The remaining gates, $g1$, $g2$, $g6$, and $g7$, are only used to buffer the latch's input and outputs from the outside world.[4]

---

[3] Note the use of the tilda '~' character to indicate the complementary output ~q. The use of tildas was discussed in Appendix A.

[4] NOTs are used in preference to BUFs, because NOTs require fewer transistors and are faster. These considerations were introduced in Chapter 6.

No-Holds-Barred Seafood Gumbo

[Editor's Note: Upon receiving and perusing the following appendix, I concluded that the strain of book authorship has taken a severe toll on Max. However . . . it's a great gumbo!]

Well, my little avocado pears, festooned with slivers of spring onions and deluged with lashings of spicy Italian dressing . . . Having absorbed the myriad juicy snippets of information which are strewn lavishly throughout *Bebop to the Boolean Boogie,* you are now uniquely qualified to regale an assembled throng with a discourse on almost any subject under the sun. So, now's the time to rest our weary brains and turn our attention to other facets of life's rich tapestry. For example, as soon as I've finished penning these last few words, I must return to practicing my unicycle skills as, I have no doubt, so do you.

But first we need to recline, relax, lay back, unwind, and take the time to recharge our batteries—and what could be more appropriate than a steaming bowl of no-holds-barred seafood gumbo? So, here's the recipe for a fulsomly flavored Epicurean taste-fest sensation sufficient to cause the most sophisticated of gastronomes to start salivating surreptitiously and to make a grown man break down and cry. This pert little beauty will pulsate promiscuously across your pallet and pummel it with a passion, titillate your taste buds and have them tap-dancing the tango on your tongue, reverberate and resonate resoundingly throughout your nervous system, and warm the cockles of your heart. In short, this frisky little number will grab you by the short-and-curlies, swing you sybaritically around the room in a syncopated symphony of delight, and leave you groveling on your knees, gnashing your teeth, and gasping for more.

The following ingredients are for the main body of the gumbo—you'll have to sort out any rice, bread, and side dishes by yourselves:

2 cups of diced onions

1½ cups of diced green bell peppers

1 cup of diced celery

2 cups of halved button mushrooms

3 large cloves of finely diced garlic

2 finely diced scotch bonnets or habañero (peppers)

10 thick-cut slices of bacon

1 pound of Cajun-style sausage

1 pound of peeled, medium-sized shrimp

½ pound of scallops

¾ pound white fish cut into slices

1 small tin anchovies

2 bay leaves

½ teaspoon of dried thyme leaves

¼ teaspoon dried oregano

½ teaspoon of salt

1½ teaspoons of white pepper

½ teaspoon of cayenne pepper

½ teaspoon of black pepper

2 teaspoons of Gumbo Filé

5½ cups of chicken stock

Lots of butter

¾ cup flour

1 bottle of dry sherry

If you can't get Cajun-style sausage, then Polish sausage will do nicely. Note that the teaspoon quantities in the list of ingredients do not refer to level measures, nor do I expect you to attempt to achieve a new world record for the amount you can balance on one spoon—just try to aim for roughly the same sensuously rounded profile you get when you're casually spooning sugar into a cup of coffee (except in the case of the cayenne pepper—try to err on the side of caution here). So, without further ado, let's gird up our loins and proceed to the fray:

1) First of all there's an art to cooking, and it starts by doing the washing up you've been putting off all day and putting all of the pots away.

2) Grill the bacon until it's crispy and crunchy; then put it on a plate to cool and set it to one side.

3) Prepare all of the vegetables, mushrooms, scotch bonnets, and garlic.

Put them all in separate bowls except for the scotch bonnets and garlic, which can go together.

4) Chop the Cajun-style sausage into ½" pieces; put them in a bowl and set it to one side.

5) Mix the salt, thyme, oregano, Gumbo Filé, and the white, black, and cayenne peppers together in a cup and set it to one side (you'll need your hands free later).

6) Wash up all the knives, chopping boards, and everything else you've used and put them all away, and then wipe down all of your working surfaces. Trust me—you'll feel better when everything is clean and tidy—have I ever lied to you? Take a five-minute break and quaff[1] a glass of wine—after all, who deserves it more than you?

7) Put the chicken stock into a large chili pan and bring it to the boil. Then reduce the heat to a low, slow simmer and leave it on the back burner.

8) Using a medium to medium-high heat, melt ¾ of a cup of butter in a large, heavy skillet until it starts to bubble. Gradually add the flour using a whisk and stir constantly until the resulting roux is a dark-ish, red-ish brown. Remove the skillet from the heat, but keep on stirring until it's cooled down enough so that the mixture won't stick to it and burn.

9) Maintain the stock at a low simmer and add the mixture that you've just made, stirring it in one spoonful at a time, and waiting for each spoonful to dissolve before adding the next.

10) Clean the skillet, put it on a medium-high to high heat, and melt a chunk of butter. Stirring all the time, sauté the celery for one minute, add the bell peppers and sauté for 1½ minutes, add the onions and sauté for 1½ minutes, then add the scotch bonnets and garlic along with the mixture of herbs, salt, and pepper and sauté for one more minute. Finally, add a slosh of sherry and keep stirring until it's all evaporated, then chuck the whole lot into the chili pan with the stock.

---

[1] Quaffing is like drinking, except that you tend to spill more down your chest.

11) Break the bacon into ½" pieces and toss them into the stock. Flake the anchovies with a fork and cast them into the stock. Hurl in the Cajun-style sausage and the bay leaves and add a healthy slosh (approximately ⅛ of a cup) of the dry sherry—you can drink the rest of the bottle later. Also, if you happen to have any lying around, add a couple of teaspoons of English Worcestershire sauce. Cover the chili pan and leave on a low simmer.

12) Return the skillet to a medium-high to high heat and melt another chunk of butter. Sauté the mushrooms until they're golden brown and squealing for more, then use them to swell the contents of the chili pan.

13) Simmer the whole mixture (stirring often) for at least one hour which, by some strange quirk of fate, will give you all the time you need to wash the skillet and the dishes you used and put them away again. If you're ravenous then you can proceed immediately to the next step but, if you're wise, you'll remove the heat and leave your cunningly captivating creation to stand overnight (chilies, stews, curries, and gumbos always taste better if the ingredients have the time to formally introduce themselves). When you're ready to chow-down, heat it back up again and proceed to the next step.

14) Add the shrimp, scallops, and fish. Bring to the boil then return to a simmer. Maintain the simmer until the seafood is cooked (I personally opt for around 15 minutes just to make sure) and you're ready to rock and roll.

This little beauty will put hairs on your chest, make them curl, and then take them off again. Seriously, this gumbo really is seductively, scintillatingly, scorchingly spicy—so if you like your dishes less on the humongously hot side, then lose one of the scotch bonnets and only use ¾ of the stated amounts of the white, black, and cayenne peppers.

You can serve your gorgeously gigantean gourmet gumbo over steamed or boiled rice, with crusty French bread, or with whatever else your hearts' desire. The quantities given above will serve eight to ten manly-man sized portions with a little something left over for the following day. Of course, no meal would be complete without some wine—and the perfect complement to your rapacious repast is to be found in a very large bottle. Enjoy!

Abbreviations and Acronyms

Electronic engineers and computer buffs simply love abbreviations and acronyms. They are really happy if they can use one abbreviation to represent multiple things or different abbreviations to represent the same things, and they become close to ecstatic if they can generate a plethora of abbreviations from permutations of the same three or four letters. The situation is further confused because the meanings behind abbreviations and acronyms may evolve over time. The list below contains the abbreviations and acronyms used in this book. For further reading, one could do a lot worse than *The Computer Glossary* by Alan Freedman.

2-D	Two Dimensional
3-D	Three Dimensional
A/D	Analog-to-Digital
ASIC	Application-Specific Integrated Circuit
ASSP	Application-Specific Standard Part
BDD	Binary Decision Diagram
BGA	Ball Grid Array
BiCMOS	Bipolar-CMOS
BiNMOS	Bipolar-NMOS
BIST	Built-In Self-Test
BJT	Bipolar Junction Transistor
CGA	Column Grid Array
CGH	Computer-Generated Hologram
CMOS	Complementary Metal-Oxide Semiconductor
COB	Chip-On-Board
COC	Chip-On-Chip

COF	Chip-On-Flex
CPLD	Complex PLD
CPU	Central Processing Unit
CRC	Cyclic Redundancy Check
CSIC	Customer-Specific Integrated Circuit
CVD	Chemical Vapor Deposition
CVI	Chemical Vapor Infiltration
D/A	Digital-to-Analog
DFT	Design-for-Test
DIL	Dual-In-Line
DRAM	Dynamic RAM
DWB	Discrete Wired Board
DSP	Digital Signal Processor
DTL	Diode-Transistor Logic
ECL	Emitter-Coupled Logic
EBE	Electron Beam Epitaxy
EDSAC	Electronic Delay Storage Automatic Calculator

EDVAC	Electronic Discrete Variable Automatic Computer
E²PROM	Electrically-Erasable Programmable Read-Only Memory
EEPROM	Electrically-Erasable Programmable Read-Only Memory
ENIAC	Electronic Numerical Integrator and Calculator
EPROM	Erasable Programmable Read-Only Memory
ESD	Electro-Static Discharge
FET	Field-Effect Transistor
FIFO	First-In First-Out
FPC	Flexible Printed Circuit
FPGA	Field-Programmable Gate Array
FPIC	Field-Programmable Interconnect Chip[1]
FPID	Field-Programmable Interconnect Device
FSM	Finite State Machine
GIGO	Garbage-In Garbage-Out
HTCC	High-Temperature Cofired Ceramics
IC	Integrated Circuit
ICR	In-Circuit Reconfigurable
I/O	Input/Output
IR	Infrared

ISP	In-System Programmable
JEDEC	Joint Electronic Device Engineering Council
LFSR	Linear Feedback Shift Register
LIFO	Last-In-First-Out
LSB	Least Significant Bit
LSB	Least Significant Byte
LSI	Large-Scale Integration
LTCC	Low-Temperature Cofired Ceramics
LTH	Lead Through-Hole
MBE	Molecular Beam Epitaxy
MCM	Multichip Module
MCM-C	Multichip Module - Ceramic substrate
MCM-D	Multichip Module - Dielectric covered substrate
MCM-L	Multichip Module - Laminate substrate
MCM-S	Multichip Module - Semiconductor substrate
MOS	Metal-Oxide Semiconductor
MSB	Most Significant Bit
MSB	Most Significant Byte
MSI	Medium-Scale Integration
NMOS	N-channel MOS
NPN	N-type - P-type - N-type
nvRAM	Non-Volatile RAM

---

[1] FPIC is a trademark of Aptix Corporation.

PAL[2]	Programmable Array Logic		**RWM**	Read-Write Memory
PCB	Printed Circuit Board		**RTL**	Resistor-Transistor Logic
PGA	Pad Grid Array		**SIPO**	Serial-In Parallel-Out
PGA	Pin Grid Array		**SISO**	Serial-In Serial-Out
PHB	Photochemical Hole Burning		**SMD**	Surface Mount Device
PISO	Parallel-In Serial-Out		**SMOBC**	Solder Mask Over Bare Copper
PLA	Programmable Logic Array		**SMT**	Surface Mount Technology
PLD	Programmable Logic Device		**SOP**	Small Outline Package
PMOS	P-channel MOS		**SRAM**	Static RAM
PNP	P-type - N-type - P-type		**SSI**	Small-Scale Integration
PROM	Programmable Read-Only Memory		**TAB**	Tape Automated Bonding
PTH	Plated Through-Hole		**TTL**	Transistor-Transistor Logic
PWB	Printed Wire Board		**ULSI**	Ultra-Large-Scale Integration
QFP	Quad Flat Pack		**UNIVAC**	Universal Automatic Computer
RAM	Random Access Memory		**ULA**	Uncommitted Logic Array
ROM	Read-Only Memory		**UV**	Ultraviolet
			VLSI	Very-Large-Scale Integration

---

[2] PAL is a registered trademark of Monolithic Memories Inc.

Glossary

Absolute Scale of Temperature - A scale of temperature which was invented by the British mathematician and physicist William Thomas, first Baron of Kelvin. Under the absolute, or Kelvin, scale of temperature, 0 K (corresponding to –273°C) is the coldest possible temperature and is known as absolute zero.

Absolute Zero - *see Absolute Scale of Temperature.*

Active-Low - Beware, here be dragons. In this book an active-low signal is defined as a signal whose active state is considered to be a logic 0. This definition, which is used by the engineers in the trenches, allows all forms of logic, including assertion-level logic, to be represented without any confusion, regardless of whether positive- or negative-logic implementations are employed.

Unfortunately, some academics (and even text books) define an active-low signal as being one whose asserted (TRUE or logic 1) state is at a lower voltage level than its unasserted (FALSE or logic 0) state. However, this definition serves only to cause confusion when combined with negative-logic implementations.

When using the term active-low in discussions with other people, you are strongly advised to make sure that you're all talking about the same thing before you start.

Active-High - Beware, here be dragons. In this book an active-high signal is defined as a signal whose active state is considered to be a logic 1. This definition, which is used by the engineers in the trenches, allows all forms of logic, including assertion-level logic, to be represented without any confusion, regardless of whether positive- or negative-logic implementations are employed.

Unfortunately, some academics (and even text books) define an active-high signal as being one whose asserted (TRUE or logic 1) state is at a higher voltage level than its unasserted (FALSE or logic 0) state. However, this definition serves only to cause confusion when combined with negative-logic implementations.

When using the term active-high in discussions with other people, you are strongly advised to make sure that you're all talking about the same thing before you start.

Active Substrate - A hybrid or multichip module substrate formed from a semiconductor. Termed active because components such as transistors can be fabricated directly into the substrate.

Active Trimming - The process of trimming components such as resistors while the circuit is under power. Such components are fabricated directly onto the substrate of a hybrid or multichip module, and the trimming is usually performed using a laser beam.

Actuator - A transducer that converts an electronic signal into a physical equivalent. For example, a loudspeaker is an actuator which converts electronic signals into corresponding sounds.

Adaptive Hardware - Refers to devices which allow new design variations to be "compiled" in real-time, which may be thought of as dynamically creating subroutines in hardware (*see also Virtual Hardware and Cache Logic*).

Additive Process - A process in which conducting material is added to specific areas of a substrate. Groups of tracks, individual tracks, or portions of tracks can be built up to precise thicknesses by iterating the process multiple times with selective masking.

Address Bus - A unidirectional set of signals used by a computer to point to memory locations in which it is interested.

A/D - *see Analog-to-Digital.*

Analog - A continuous value that most closely resembles the real world and can be as precise as the measuring technique allows.

Analog Circuit - A collection of components used to process or generate analog signals.

Analog-to-Digital (A/D) - The process of converting an analog value into a digital equivalent.

Analogue - The way they spell "analog" in England.

Anisotropic Adhesive - Special adhesives which contain minute particles of conductive material. These adhesives find particular application with the flipped-chip techniques used to mount bare die on the substrates of hybrids, multichip modules, or circuit boards. The conducting particles are only brought in contact with each other at the sites where the raised pads on the die are pressed down over their corresponding pads on the substrate, thereby forming good electrical connections between the pads.

Anti-Fuse Technology - A programmable logic device technology in which conducting paths (anti-fuses) are grown by applying signals of relatively high voltage and current to the device's inputs.

Anti-Pad - The area of copper etched away around a via or a plated through-hole on a power or ground plane, thereby preventing an electrical connection being made to that plane.

Application-Specific Integrated Circuit (ASIC) - A device whose function is determined by a designer for a particular application or group of applications.

Application-Specific Standard Part (ASSP) - Refers to an integrated circuit created by a device manufacturer using ASIC technologies, and for these components to be sold as standard parts to anybody who wants to buy them.

ASIC - *see Application-Specific Integrated Circuit.*

ASIC Cell - A logic function in the cell library defined by the manufacturer of an application-specific integrated circuit.

Assertion-Level Logic - A technique used to more precisely represent the function of logic gates with active-low inputs.

Associative Rules - Algebraic rules that state that the order in which pairs of variables are associated together will not affect the result of an operation; for example,[1] $(a \,\&\, b) \,\&\, c \equiv a \,\&\, (b \,\&\, c)$.

ASSP - *see Application-Specific Standard Part.*

Asynchronous - A signal whose data is acknowledged or acted upon immediately, regardless of any clock signal.

Atlo - Unit qualifier (symbol = a) representing one millionth of one millionth of one millionth, or 10^{-18}. For example, 3aS stands for 3×10^{-18} seconds.

Backplane - The medium used to interconnect a number of circuit boards. Typically refers to a special, heavy-duty printed or discrete wired circuit board.

Ball Grid Array (BGA) - A packaging technology similar to a pad grid array, in which a device's external connections are arranged as an array of conducting pads on the base of the package. However, in the case of a ball grid array, small balls of solder are attached to the conducting pads.

Bare Die - An unpackaged integrated circuit.

Barrier Layer - *see Overglassing.*

Base - Refers to the number of digits in a numbering system. For example, the decimal numbering system is said to be base-10. May also be referred to as the "radix."

Basic Cell - A predefined group of unconnected components that is replicated across the surface of a gate array.

BDD - *see Binary Decision Diagram.*

Bebop - A form of music characterized by fast tempos and agitated rhythms that became highly popular in the decade following World War II.

BiCMOS - A technology in which the function of each logic gate is implemented using low-power CMOS, while the output stage is implemented using high-drive bipolar transistors.

Binary - Base-2 numbering system. The type of numbers that computers use internally.

Binary Decision Diagram (BDD) - A method of representing Boolean equations as decision trees.

Binary Digit - A numeral in the binary scale of notation. A binary digit can adopt one of two values: 0 or 1. Often abbreviated to "bit."

Binary Encoding - A form of state assignment for state machines that requires the minimum number of state variables.

Binary Logic - Digital logic gates based on two distinct voltage levels. The two voltages are used to represent the binary values 0 and 1, and their logical equivalents FALSE and TRUE.

BiNMOS - A relatively new low-voltage integrated circuit technology in which complex combinations of bipolar and NMOS transistors are used to form sophisticated output stages providing both high speed and low static power dissipation.

---

[1] Note that the symbol $\equiv$ indicates *"is equivalent to"* or *"is the same as."*

Bipolar Junction Transistor (BJT) - A family of transistors.

Bi-quinary - A system which utilizes two bases, base-2 and base-5, to represent decimal numbers. Each decimal digit is represented by the sum of two parts, one of which has the value of decimal zero or five, and the other the values of zero through four. The abacus is one practical example of the use of a bi-quinary system.

BIST - *see Built-In Self-Test.*

Bit - Abbreviation of *binary digit*. A binary digit can adopt one of two values: 0 or 1.

BJT - *see Bipolar Junction Transistor.*

Blind Via - A via that is only visible from one side of the substrate.

Bobble - A small circle used on the inputs to a logic gate symbol to indicate an active-low input or control, or on the outputs to indicate a negation (inversion) or a complementary signal. Some engineers prefer to use the term bubble.

Boolean Algebra - A mathematical way of representing logical expressions.

Boot-Strapping - A sequence of initialization operations performed by a computer when it is first powered up.

Bounce Pad - A special pattern etched onto the power or ground plane of a micro-wire circuit board to be used in conjunction with a laser beam which is employed to create blind vias. The laser beam evaporates the epoxy forming the outer layers of the board and continues down to the bounce pad which reflects, or bounces, it back up, thereby terminating the via.

Braze - To unite or fuse two pieces of metal by heating, or with a hard solder with a high melting point.

Built-In Self-Test (BIST) - A test strategy in which additional logic is built into a component, thereby allowing it to test itself.

Bulk Storage - Refers to some form of media, typically magnetic, such as tape or a disk drive which can be used to store large quantities of information relatively inexpensively.

Buried Via - A via used to link conducting layers internal to a substrate. Such a via is not visible from either side of the substrate.

Bus - A set of signals performing a common function and carrying similar data. Typically represented using vector notation; for example, address[7:0].

Bundle - A set of signals related in some way that makes it appropriate to group them together for ease of representation or manipulation. May contain both scalar and vector elements; for example, {a,b,c,d[5:0]}.

Byte - A group of eight binary digits, or bits.

Cache Logic[2] - The proprietary name intended to reflect designs realized in virtual hardware. Derived from its similarity to the concept of *Cache Memory*.

---

[2] Cache Logic is a trademark of Atmel Corporation.

Cache Memory - A small, high-speed, memory (usually SRAM) used to buffer the central processing unit from any slower, lower cost memory devices such as DRAM. The high-speed cache memory is used to store active data,[3] while the bulk of the data resides in the slower memory.

Canonical Form - In a mathematical context this term is taken to mean a generic or basic representation. Canonical forms provide the means to compare two expressions without falling into the trap of trying to compare "apples" with "oranges."

Capacitance - A measure of the ability of two adjacent conductors separated by an insulator to hold a charge when a voltage differential is applied between them. Capacitance is measured in units of Farads.

Catalyst - A substance that initiates a chemical reaction under different conditions (such as lower temperatures) than would otherwise be possible. The catalyst itself remains unchanged at the end of the reaction.

Cell - *see ASIC Cell, Basic Cell, Cell Library, and Memory Cell.*

Cell Library - The collective name for the set of logic functions defined by the manufacturer of an application-specific integrated circuit. The designer decides which types of cells should be realized and connected together to make the device perform its desired function.

Central Processing Unit (CPU) - The "brain" of a computer where all of the number-crunching is performed.

Ceramic - An inorganic, nonmetallic material, such as alumina, beryllia, steatite, or forsterite, which is fired at a high temperature and is often used in electronics as a substrate or to create component packages.

CGA - *see Column Grid Array.*

CGH - *see Computer-Generated Hologram.*

Channel - **(1)** The area between two arrays of basic cells in a channeled gate array. **(2)** The gap between the source and drain regions in a MOS transistor.

Channeled Gate Array - Application-specific integrated circuit organized as arrays of basic cells. The areas between the arrays are known as channels.

Channel-Less Gate Array - Application-specific integrated circuit organized as a single large array of basic cells. May also be referred to as a "sea of cells" or a "sea of gates" device.

Checksum - The final cyclic-redundancy-check value stored in a linear feedback shift register (or software equivalent). Also known as a "signature" in the guided-probe variant of functional test.

Chemically-Amplified Resist - In the case of a chemically-amplified resist, the application of a relatively small quantity of

---

[3] In this context, "*active data*" refers to data or instructions that a program is currently using, or which the operating system believes that the program will want to use in the immediate future.

ultraviolet light stimulates the formation of chemicals in the resist which accelerates the degrading process. This reduces the amount of ultraviolet light which is required to degrade the resist and allows the creation of finer features with improved accuracy.

Chemical Vapor Deposition (CVD) - A process for growing thin films on a substrate, in which a gas containing the required molecules is converted into a plasma by heating it to extremely high temperatures using microwaves. The plasma carries atoms to the surface of the substrate where they are attracted to the crystalline structure of the substrate. This underlying structure acts as a template. The new atoms continue to develop the structure to build up a layer on the substrate's surface.

Chemical Vapor Infiltration (CVI) - A process similar to chemical vapor deposition (CVD) but, in this case, the process commences by placing a crystalline powder of the required substance in a mold. Additionally, thin posts, or columns, can be preformed in the mold, and the powder can be deposited around them. When exposed to the same plasma as used in the CVD technique, the powder coalesces into a polycrystalline mass. After the CVI process has been performed, the posts can be dissolved leaving holes through the crystal for use in creating vias. CVI processes can produce layers twice the thickness of those obtained using CVD techniques at a fraction of the cost.

Chip - Popular name for an integrated circuit.

Chip-On-Board (COB) - A process in which unpackaged integrated circuits are physically and electrically attached to a circuit board, and are then encapsulated with a "glob" of protective material such as epoxy.

Chip-On-Chip (COC) - A process in which unpackaged integrated circuits are mounted on top of each other. Each die is very thin and it is possible to have over a hundred die forming a 3-D cube.

Chip-On-Flex (COF) - Similar to chip-on-board, except that the unpackaged integrated circuits are attached to a flexible printed circuit.

Circuit Board - The generic name for a wide variety of interconnection techniques, which include rigid, flexible, and rigid-flex boards in single-sided, double-sided, multilayer, and discrete wired configurations.

CMOS - Logic gates constructed using both NMOS and PMOS transistors connected in a complementary manner.

Coaxial Cable - A conductor in the form of a central wire surrounded first by a dielectric (insulating) layer, and then by a conducting tube which serves to shield the central wire from external interference.

COB - *see Chip-On-Board.*

COC - *see Chip-On-Chip.*

Coefficient of Thermal Expansion - Defines the amount a material expands and contracts due to changes in temperature. If materials with different coefficients of thermal expansion are bonded together, changes in temperature will cause shear forces at the interface between them.

COF - *see Chip-On-Flex.*

Cofired Ceramic - A substrate formed from multiple layers of "green" ceramic that are bonded together and fired at the same time.

Column Grid Array (CGA) - A packaging technology similar to a pad grid array, in which a device's external connections are arranged as an array of conducting pads on the base of the package. However, in the case of a column grid array, small columns of solder are attached to the conducting pads.

Combinatorial - *see Combinational.*

Combinational - A digital function whose output value is directly related to the current combination of values on its inputs.

Commutative Rules - Algebraic rules that state that the order in which variables are specified will not affect the result of an operation; for example,[4] $a \& b \equiv b \& a$.

Comparator (digital) - A logic function that compares two binary values, and outputs the results in terms of binary signals representing *less-than* and/or *equal-to* and/or *greater-than.*

Compiled Cell Technology - A technique used to create portions of a standard cell application-specific integrated circuit. The masks used to create components and interconnections are directly generated from Boolean representations using a silicon compiler. May also be used to create data-path functions and memory functions.

Complementary Output - Refers to a function with two outputs carrying complementary logical values. One output is referred to as the *true output* and the other as the *complementary output.*

Complementary Rules - Rules in Boolean Algebra derived from the combination of a single variable with the inverse of itself.

Complex Programmable Logic Device (CPLD) - Device that contains a number of PLA or PAL functions sharing a common programmable interconnection matrix.

Computer-Generated Hologram (CGH) - Refers to a slice of quartz or similar material into which three-dimensional patterns are cut using a laser. The angles of the patterns cut into the quartz are precisely calculated for use in the optical communication strategy known as holographic interconnect. All of these calculations are performed by a computer, and the laser used to cut the three-dimensional patterns into the quartz is also controlled by a computer. Thus, the slice of quartz is referred to a computer-generated hologram.

Computer Virus - There are many different types of computer viruses but, for the purposes of this book, a computer virus may be defined as a self-replicating program released into a computer system for a number of purposes. These purposes range from the simply mischievous, such as displaying humorous or annoying messages, to the downright nefarious, such as corrupting data or destroying (or subverting) the operating system.

---

[4] Note that the symbol $\equiv$ indicates *"is equivalent to"* or *"is the same as."*

Conditioning - *see Signal Conditioning.*

Conductive Ink Technology - A technique in which tracks are screen printed directly onto the surface of a circuit board using a conductive ink.

Conjunction - Propositions combined with an AND operator; for example, *"You have a parrot on your head AND you have a fish in your ear."* The result of a conjunction is true if all the propositions comprising that conjunction are true.

Configurable Hardware - A product whose function may be customized once or a very few times (*see also Reconfigurable Hardware, Remotely Reconfigurable Hardware, Dynamically Reconfigurable Hardware, and Virtual Hardware*).

CPLD - *see Complex Programmable Logic Device.*

CPU - *see Central Processing Unit.*

CRC - *see Cyclic Redundancy Check.*

Crossover - Used in discrete wire technology to designate the intersection of two or more insulated wires.

CSIC - *see Customer-Specific Integrated Circuit.*

Cure - To harden a material using heat, ultraviolet light, or some other process.

Cured - Refers to a material that has been hardened using heat, ultraviolet light, or some other process.

Customer-Specific Integrated Circuit (CSIC) - An alternative and possibly more

accurate name for an ASIC, but this term is rarely used in the industry and shows little indication of finding favor with the masses.

CVD - *see Chemical Vapor Deposition.*

CVI - *see Chemical Vapor Infiltration.*

Cyclic Redundancy Check (CRC) - A calculation used to detect errors in data communications, typically performed using a linear feedback shift register. Similar calculations may be used for a variety of other purposes such as data compression.

D/A - *see Digital-to-Analog.*

Data Bus - A bi-directional set of signals used by a computer to convey information from a memory location to the central processing unit and vice versa. More generally, a set of signals used to convey data between digital functions.

Data-Path Function - A well-defined function such as an adder, counter, or multiplier used to process digital data.

Daughter Board - If a backplane contains active components such as integrated circuits, then it is usually referred to as a mother board. In this case, the boards plugged into it are referred to as daughter boards.

Decimal - Base-10 numbering system.

Decoder (digital) - A logic function that uses a binary value, or address, to select between a number of outputs and to assert the selected output by placing it in its active state.

Deep Sub-Micron - Typically taken to refer to integrated circuits containing structures which are smaller than 0.5 microns.

Delamination - Occurs when a composite material formed from a number of layers is stressed, thermally or otherwise, such that the layers begin to separate.

DeMorgan Transformation - The transformation of a Boolean expression into an alternate, and often more convenient, form.

Die - **(1)** An unpackaged integrated circuit. In this case, the plural of die is also die (in much the same way that "a shoal of herring" is the plural of "herring"). **(2)** A piece of metal with a design engraved or embossed on it for stamping onto another material, upon which the design appears in relief.

Dielectric Layer - **(1)** An insulating layer used to separate two signal layers. **(2)** An insulating layer used to modify the electrical characteristics of an MCM-D substrate.

Die Separation - The process of separating individual die from the wafer by marking the wafer with a diamond scribe and fracturing it along the scribed lines.

Die Stacking - A technique used in specialist applications in which several bare die are stacked on top of each other to form a sandwich. The die are connected together and then packaged as a single entity.

Diffusion Layer - The surface layer of a piece of semiconductor into which impurities are diffused to form P-type and N-type material. In addition to forming components, the diffusion layer may also be used to create embedded traces.

Digital - A value represented as being in one of a finite number of discrete states called quanta. The accuracy of a digital value is dependent on the number of quanta used to represent it.

Digital Circuit - A collection of logic gates used to process or generate digital signals.

Digital Signal Processor (DSP) - A primarily digital component used to process either digital or analog signals. In the case of the latter, the signal may first be conditioned, then converted into a digital equivalent using an analog-to-digital (A/D) converter function. The signal conditioning and A/D functions may either be external to the DSP or resident in the device. A typical DSP application might be the compression/decompression of video data.

Digital-to-Analog (D/A) - The process of converting a digital value into an analog equivalent.

Diode - A two-terminal device that only conducts electricity in one direction; in the other direction it behaves like an open switch. The term diode is typically taken to refer to a semiconductor device, although alternative implementations such as vacuum tubes are available.

Diode-Transistor Logic (DTL) - Logic gates implemented using particular configurations of diodes and bipolar junction transistors. For the majority of today's designers, diode-transistor logic is of historical interest only.

Discrete Device - Typically taken to refer to an electronic component such as a resistor, capacitor, diode, or transistor that is presented in an individual package. More rarely, the term may be used in connection with a simple integrated circuit containing a small number of primitive gates.

Discrete Wire Board (DWB) - A form of circuit board in which a special computer-controlled wiring machine ultrasonically bonds extremely fine insulated wires into the surface layer of the board. This discipline has enjoyed only limited recognition, but may be poised to emerge as the technology-of-choice for high-speed designers.

Discrete Wire Technology - The technology used to fabricate discrete wire boards.

Disjunction - Propositions combined with an OR operator; for example, *"You have a parrot on your head OR you have a fish in your ear."* The result of a disjunction is true if at least one of the propositions comprising that disjunction is true.

Distributive Rules - Two very important rules in Boolean Algebra. The first states that the AND operator distributes over the OR operator; for example,[5] $a \& (b \mid c) \equiv (a \& b) \mid (a \& c)$. The second states that the OR operator distributes over the AND operator; for example, $a \mid (b \& c) \equiv (a \mid b) \& (a \mid c)$

Doping - The process of inserting selected impurities into a semiconductor to create P-type or N-type material.

Double-Sided - A printed circuit board with tracks on both sides.

DRAM - *see Dynamic RAM.*

DSP - *see Digital Signal Processor.*

DTL - *see Diode-Transistor Logic.*

Duo-decimal - Base-12 numbering system.

DWB - *see Discrete Wired Board.*

Dynamic Flex - A type of flexible printed circuit which is used in applications that are required to undergo constant flexing such as ribbon cables in printers.

Dynamically Reconfigurable Hardware - A product whose function may be customized on-the-fly while remaining resident in the system (*see also Configurable Hardware, Reconfigurable Hardware, Remotely Reconfigurable Hardware, and Virtual Hardware*).

Dynamic RAM (DRAM) - A memory device in which each cell is formed from a transistor-capacitor pair. Called dynamic because the capacitor loses its charge over time, and each cell must be periodically recharged if it is to retain its data.

EBE - *see Electron Beam Epitaxy.*

ECL - *see Emitter-Coupled Logic.*

Edge-Sensitive - An input that only affects a function when it transitions from one logic value to another.

E²PROM - *see Electrically-Erasable Programmable Read-Only Memory.*

EEPROM - *see Electrically-Erasable Programmable Read-Only Memory.*

Electrically-Erasable Programmable Read-Only Memory (EEPROM) - A memory device whose contents can be electrically programmed by the designer. Additionally,

---

[5] Note that the symbol $\equiv$ indicates *"is equivalent to"* or *"is the same as."*

the contents can be electrically erased, allowing the device to be reprogrammed.

Electromigration - **(1)** A process in which structures on an integrated circuit's substrate[6] are eroded by the flow of electrons in much the same way as land is eroded by a river. **(2)** The process of forming transistor-like regions in a semiconductor using an intense magnetic field.

Electron Beam Epitaxy (EBE) - A technique for creating thin films on substrates in precise patterns, in which the substrate is first coated with a layer of dopant material before being placed in a high vacuum. A guided beam of electrons is fired at the substrate, causing the dopant to be driven into it, effectively allowing molecular-thin layers to be "painted" onto the substrate where required.

Electron Beam Lithography - An integrated circuit fabrication process in which fine beams of electrons are used to draw extremely high-resolution patterns directly into the resist without the use of a mask.

Electro-Static Discharge (ESD) - The process of moving around can generate static electricity. The term electro-static discharge refers to a charged person, or object, discharging static electricity. Although the current associated with such a static charge is low, the electric potential can be in the millions of volts and can severely damage electronic components. CMOS devices are particularly prone to damage from static electricity.

Emitter-Coupled Logic (ECL) - Logic gates implemented using particular configurations of bipolar junction transistors.

Enzyme - One of numerous complex proteins which are produced by living cells and catalyze biochemical reactions at body temperatures.

EPROM - *see Erasable Programmable Read-Only Memory.*

Equivalent Gate - A concept in which each type of logic function is assigned an equivalent gate value for the purposes of comparing functions and devices. However, the definition of an equivalent gate varies depending on who you're talking to.

Equivalent Integrated Circuit - A concept used to compare the component density supported by diverse interconnection technologies such as circuit boards, hybrids, and multichip modules.

Erasable Programmable Read-Only Memory (EPROM) - A memory device whose contents can be electrically programmed by the designer. Additionally, the contents can be erased by exposing the die to ultraviolet light through a quartz window mounted in the top of the component's package.

ESD - *see Electro-Static Discharge.*

Etching - The process of selectively removing any material not protected by a resist using an appropriate solvent or acid. In some cases the unwanted material is removed using an electrolytic process.

---

[6] Particularly structures in deep sub-micron technologies.

Eutectic Bond - A bond formed when two pieces of metal, or metal-coated materials, are pressed together and vibrated at ultrasonic frequencies.

Face-Surface Mirror - A mirror in which the reflective coating is located on the face of the mirror, and not behind it as would be the case in the mirror you look into when you shave or apply your makeup (or both depending on how liberated you are).

Falling-Edge - *see Negative-Edge.*

Fan-In Via - *see Fan-Out Via.*

Fan-Out Via - In the case of surface mount devices attached to double-sided or multilayer boards, each component pad is usually connected by a short length of track to a via which forms a link to other conducting layers, and this via is known as a fan-out via. Some engineers attempt to differentiate vias that fall inside the device's footprint (under the body of the device) from vias that fall outside the device's footprint by referring to the former as *fan-in vias*, but this is not a widely used term.

Femto - Unit qualifier (symbol = f) representing one thousandth of one millionth of one millionth, or 10^{-15}. For example, 3fS stands for 3×10^{-15} seconds.

FET - *see Field-Effect Transistor.*

Field-Effect Transistor (FET) - A transistor whose control, or gate, signal creates an electro-magnetic field which turns the transistor ON or OFF.

Field-Programmable Gate Array (FPGA) - A programmable logic device which is more versatile than traditional programmable devices such as PALs and PLAs, but less versatile than an application-specific integrated circuit. Some field-programmable gate arrays use fuses such as those found in programmable logic devices, but others are based on SRAM equivalents.

Field-Programmable Interconnect Chip (FPIC)[7] - An alternative, proprietary name for a field-programmable interconnect device.

Field-Programmable Interconnect Device (FPID) - A device which is used to connect logic devices together, and which can be dynamically reconfigured in the same way as standard SRAM-based FPGAs. Because each FPID may have around 1,000 pins, only a few such devices are typically required on a circuit board

FIFO - *see First-In First-Out.*

Finite State Machine - *see State Machine.*

Firmware - Refers to programs, or sequences of instructions, that are hard-coded into non-volatile memory devices.

First-In First Out (FIFO) - A memory device in which data is read out in the same order that it was written in.

Flash - *see Gold Flash.*

FLASH Memory - An evolutionary technology that combines the best features of the EPROM and E^2PROM technologies. The name FLASH is derived from the technology's fast reprogramming time compared to EPROM.

---

[7] FPIC is a trademark of Aptix Corporation.

Flex - *see Flexible Printed Circuit.*

Flexible Printed Circuit (FPC) - A specialist circuit board technology, often abbreviated to "flex," in which tracks are printed onto flexible materials. There are a number of flavors of flex, including static flex, dynamic flex, and rigid flex.

Flipped Chip - A generic name for processes in which unpackaged integrated circuits are mounted directly onto a substrate with their component-sides facing the substrate.

Flipped TAB - A combination of flipped chip and tape automated bonding.

Footprint - The area occupied by a device mounted on a substrate.

FPC - *see Flexible Printed Circuit.*

FPGA - *see Field-Programmable Gate Array.*

FPIC - *see Field-Programmable Interconnect Chip.*

FPID - *see Field-Programmable Interconnect Device.*

FR4 - The most commonly used insulating base material for circuit boards. FR4 is made from woven glass fibers which are bonded together with an epoxy. The board is cured using a combination of temperature and pressure which causes the glass fibers to melt and bond together, thereby giving the board strength and rigidity. The first two characters stand for "Flame Retardant." FR4 is technically a form of fiberglass, and some people do refer to these composites as fiberglass boards or fiberglass substrates, but not often.

Free-Space Optical Interconnect - A form of optical interconnect in which laser-diode transmitters communicate directly with photo-transistor receivers without employing optical fibers or optical waveguides.

Full Custom - An application-specific integrated circuit in which the designer has complete control over every mask layer used to fabricate the device. The manufacturer does not provide a cell library or prefabricate any components on the substrate.

Functional Latency - Refers to the fact that, at any given time, only a portion of the logic functions in a device or system are typically active (doing anything useful).

Functional Test - A test strategy in which signals are applied to a circuit's inputs, and the resulting signals which are observed on the circuit's outputs are compared to known good values.

Fuse - *see Fusible-Link Technology.*

Fusible-Link Technology - A programmable logic device technology which employs links called fuses. Individual fuses can be removed by applying pulses of relatively high voltage and current to the device's inputs.

Fuzz-Button - A small ball of fibrous gold used in one technique for attaching components such as multichip modules to circuit boards. Fuzz-buttons are inserted between the pads on the base of the package and their corresponding pads on the board. When the package is forced against the board, the fuzz-buttons compress to form good electrical connections. Even when the pressure is removed, the fuzz-buttons act in a similar manner to Velcro™ and continue to hold the component in place. One of the main advantages of the fuzz-button approach is that it allows broken devices to be quickly removed and replaced. Even though fuzz-button technology would appear to be inherently unreliable, it is used in such devices as missiles, so one can only assume that it is fairly robust.

GaAs - *see Gallium Arsenide.*

Gallium Arsenide (GaAs) - A 3:5 valence high-speed semiconductor formed from a mixture of gallium and arsenic. GaAs transistors are approximately eight times faster than their silicon equivalents and use approximately one tenth of the power, but the material is difficult to manufacture and to work with.

Garbage-In Garbage-Out (GIGO) - An electronic engineer's joke, also familiar to the writers of computer programs.

Gate Array - An application-specific integrated circuit in which the manufacturer prefabricates devices containing arrays of unconnected components organized in groups called basic cells. The designer specifies the function of the device in terms of cells from the cell library and the connections between them, and the manufacturer then generates the masks used to create the metallization layers.

Geometry - Refers to the size of structures created on an integrated circuit. The structures typically referenced are the width of the tracks and the length of the transistor's channels; the dimensions of other features are derived as ratios of these structures.

Giga - Unit qualifier (symbol = G) representing one thousand million, or 10^9. For example, 3GHz stands for 3×10^9 Hertz.

GIGO - *see Garbage-In Garbage-Out.*

Glue Logic - Simple logic gates used to interface more complex functions together.

Gold Flash - An extremely thin layer of gold with a thickness measured on the molecular level which is either electroplated or chemically plated[8] onto a surface.

Gray Code - A sequence of binary values in which each pair of adjacent values differs by only a single bit; for example, 00, 01, 11, 10.

Green Ceramic - Unfired, malleable ceramic.

Ground Plane - A conducting layer in, or on, a substrate providing a grounding, or reference, point for components. There may be several ground planes separated by insulating layers.

Guard Condition - A Boolean expression associated with a state transition in a state diagram or state table. The expression must be satisfied for that state transition to be executed.

Guided Probe - A form of functional test in which the operator is guided in the probing of a circuit to isolate a faulty component or track.

Guided-Wave - A form of optical interconnect, in which optical waveguides are fabricated directly on the substrate of a

---

8 That is, using an electroless plating process.

multichip module. These waveguides can be created using variations on standard opto-lithographic thin-film processes.

Hard Macro (Macro Cell) - A logic function defined by the manufacturer of an application-specific integrated circuit. The function is described in terms of the simple functions provided in the cell library and the connections between them. The manufacturer also defines how the cells forming the macro will be assigned to basic cells and the routing of tracks between the basic cells.

Hardware - Generally understood to refer to any of the physical portions constituting an electronic system, including components, circuit boards, power supplies, cabinets, and monitors.

Hertz - Unit of frequency. One Hertz equals one cycle, or one oscillation, per second.

Heterojunction - The interface between two regions of dissimilar semiconductor materials. The interface of a heterojunction has naturally occurring electric fields which can be used to accelerate electrons, and transistors created using heterojunctions can switch much faster than their homo-junction counterparts of the same size.

Hexadecimal - Base-16 numbering system. Each hexadecimal digit can be directly mapped onto four binary digits, or bits.

High Impedance State - The state on a signal that is not being driven by any value. A high-impedance state is indicated by the character Z.

Hologram - A three-dimensional image (from the Greek: *holos* meaning "whole" and *gram* meaning "message").

Holographic Interconnect - A form of optical interconnect based on a thin slice of quartz, into which three-dimensional images are cut using a laser beam. Thus, the quartz is referred to as a computer-generated hologram, and this interconnection strategy is referred to as holographic.

Holography - The art of obtaining three-dimensional images known as holograms.

Homojunction - An interface between two regions of semiconductor having the same basic composition but opposing types of doping. Homojunctions dominate current processes because they are easier to fabricate than heterojunctions.

Hybrid - An electronic sub-system in which a number of integrated circuits (packaged and/or unpackaged) and discrete components are attached directly to a common substrate. Connections between the components are formed on the surface of the substrate, and some components such as resistors and inductors may also be fabricated directly onto the substrate.

Hydrogen Bond - The electrons in a water molecule are not distributed equally, because the oxygen atom is a bigger, more robust fellow which grabs more than its fair share. The end result is that the oxygen atom has an overall negative charge, while the two hydrogen atoms are left feeling somewhat on the positive side. This unequal distribution of charge means that the hydrogen atoms are attracted to anything with a negative bias; for example, the oxygen atom of another water molecule. The resulting bond is known as a hydrogen bond.

Hz - *see Hertz*.

IC - *see Integrated Circuit*.

ICR - *see In-Circuit Reconfigurable*.

Idempotent Rules - Rules derived from the combination of a single Boolean variable with itself.

Impedance - The resistance to the flow of current caused by resistive, capacitive, or inductive devices (or undesired elements) in a circuit.

In-Circuit Reconfigurable (ICR) - An SRAM-based, or similar component which can be dynamically reprogrammed on-the-fly while remaining resident in the system.

Inclusion - A defect in a crystalline structure.

Inductance - A property of a conductor that allows is to store energy in a magnetic field which is induced by a current flowing through it. Inductance is measured in units of Henries (the base unit is a Henry).

Inert Gas - *see Noble Gas*.

In-System Programmable (ISP) - An E^2-based, FLASH-based, or similar component which can be reprogrammed while remaining resident on the circuit board.

Integrated Circuit (IC) - A device in which components such as resistors, capacitors, diodes, and transistors are formed on the surface of a single piece of semiconductor.

Invar - An alloy similar to bronze.

Involution Rule - A rule that states that an even number of Boolean inversions cancel each other out.

Ion - A particle formed when an electron is added to, or subtracted from, a neutral atom or group of atoms.

Ion Implantation - A process in which beams of ions are directed at a semiconductor to alter its type and conductivity in certain regions.

ISP - *see In-System Programmable*.

JEDEC - *see Joint Electronic Device Engineering Council*.

Jelly Bean Device - An integrated circuit containing a small number of simple logic functions.

Joint Electronic Device Engineering Council (JEDEC) - A council which creates, approves, arbitrates, and oversees industry standards for electronic devices. In programmable logic, the term JEDEC refers to a textual file containing information used to program a device. The file format is a JEDEC approved standard and is commonly referred to as a JEDEC file.

Jumper - A small piece of wire used to link two tracks on a circuit board.

Karnaugh Map - A graphical technique for representing a logical function. Karnaugh maps are often useful for the purposes of minimization.

Kelvin Scale of Temperature - A scale of temperature which was invented by the

British mathematician and physicist William Thomas, first Baron of Kelvin. Under the Kelvin, or absolute, scale of temperature, 0 K (corresponding to –273°C) is the coldest possible temperature and is known as absolute zero.

Kilo - Unit qualifier (symbol = K) representing one thousand, or 10^3. For example, 3KHz stands for 3×10^3 Hertz.

Kipper - A fish cured by smoking and salting.[9]

Latch-Up Condition - A condition in which a circuit draws uncontrolled amounts of current, and certain voltages are forced, or "latched-up," to some level. Particularly relevant in the case of CMOS devices which can latch-up if their operating conditions are violated.

Laminate - A material constructed from thin layers or sheets. Often used in the context of circuit boards.

Large-Scale Integration (LSI) - Refers to the number of logic gates in a device. By one convention, large-scale integration represents a device containing 100 to 999 gates.

Laser Diode - A special semiconductor diode which emits a beam of coherent light.

Last-In First Out (LIFO) - A memory device in which data is read out in the reverse order to which it was written in.

Lateral Thermal Conductivity - Good lateral thermal conductivity means that the heat generated by components mounted on a substrate can be conducted horizontally across the substrate and out through its leads.

Lead - **(1)** A metallic element (chemical symbol Pb). **(2)** A metal conductor used to provide a connection from the inside of a device package to the outside world for soldering or other mounting techniques. Leads are also commonly called pins.

Lead Frame - A metallic frame containing leads and a base to which an unpackaged integrated circuit is attached. After encapsulation, the outer part of the frame is cut away and the leads are bent into the required shapes.

Lead Through-Hole (LTH) - A technique for populating circuit boards in which component leads are inserted into plated through-holes. Often abbreviated to "through-hole" or "thru-hole." When all of the components have been inserted, they are soldered to the board, usually using a wave soldering technique.

Least-Significant Bit (LSB) - The binary digit, or bit, in a binary number that represents the least significant value (typically the right-hand bit).

Least-Significant Byte (LSB) - The byte in a multi-byte word that represents the least significant values (typically the right-hand byte).

Level-Sensitive - An input whose effect on a function depends only on its current logic value or level, and is not directly related to it transitioning from one logic value to another.

---

[9] Particularly tasty as a breakfast dish.

LFSR - *see Linear Feedback Shift Register.*

LIFO - *see Last-In First-Out.*

Line - Used to refer to the width of a track; for example, *"This track has a line-width of 0.12 mm."*

Linear Feedback Shift Register (LFSR) - A shift register whose data input is generated as an XOR or XNOR of two or more elements in the register chain.

Literal - A variable (either true or inverted) in a Boolean equation.

Low-Fired Cofired - Similar in principle to standard cofired ceramic substrate techniques. However, low-fired cofired uses modern ceramic materials with compositions that allow them to be fired at temperatures as low as 850°C. Firing at these temperatures in an inert atmosphere such as nitrogen allows non-refractory metals such as copper to be used to create tracks.

Logic Function - A mathematical function that performs a digital operation on digital data and returns a digital value.

Logic Gate - The physical implementation of a logic function.

Logic Synthesis - A process in which a program is used to optimize the logic used to implement a design.

LSB - *see Least-Significant Bit and Least-Significant Byte.*

LSI - *see Large-Scale Integration.*

LTH - *see Lead Through-Hole.*

Macro Cell - *see Hard Macro.*

Macro Function - *see Soft Macro.*

Mask - *see Optical Mask.*

Mask Programmable - A device such as a read-only memory which is programmed during its construction using a unique set of masks.

Maximal Displacement - A linear feedback shift register whose taps are selected such that changing a single bit in the input data stream will cause the maximum possible disruption to the register's contents.

Maximal Length - A linear feedback shift register that sequences through $(2^n - 1)$ states before returning to its original value.

Max's Algorithm - A never-before-seen-in-public, soon-to-be-famous, mind-bogglingly-convoluted algorithm for automatically extracting Reed-Müller expressions.

Maxterm - The logical OR of the inverted variables associated with an input combination to a logical function.

MBE - *see Molecular Beam Epitaxy.*

MCM - *see Multichip Module.*

Medium-Scale Integration (MSI) - Refers to the number of logic gates in a device. By one convention, medium-scale integration represents a device containing 13 to 99 gates.

Mega - Unit qualifier (symbol = M) representing one million, or 10^6. For example, 3MHz stands for 3×10^6 Hertz.

Memory Cell - A unit of memory used to store a single binary digit, or bit, of data.

Memory Word - A number of memory cells logically and physically grouped together. All the cells in a word are typically written to, or read from, at the same time.

Metallization Layer - A layer of conducting material on an integrated circuit that is selectively deposited or etched to form connections between logic gates. There may be several metallization layers separated by dielectric (insulating) layers.

Metal-Oxide Semiconductor (MOS) - A family of transistors.

Meta-Stable - A condition where the outputs of a logic function are oscillating uncontrollably between undefined values.

Micro - Unit qualifier (symbol = μ) representing one millionth, or 10^{-6}. For example, 3μS stands for 3×10^{-6} seconds.

Microwire[10] - A trade name for one incarnation of discrete wire technology. Microwire augments the main attributes of multiwire with laser-drilled blind vias, allowing these boards to support the maximum number of tracks and components.

Milli - Unit qualifier (symbol = m) representing one thousandth, or 10^{-3}. For example, 3mS stands for 3×10^{-3} seconds.

Minimization - The process of reducing the complexity of a Boolean expression.

Minterm - The logical AND of the variables associated with an input combination to a logical function.

Mod - *see Modulus.*

Modulus - Refers to the number of states that a function such as a counter will pass through before returning to its original value. For example, a function that counts from 0000_2 to 1111_2 has a modulus of 16 and would be called a modulo-16 or mod-16 counter.

Modulo - *see Modulus.*

Molecular Beam Epitaxy (MBE) - A technique for creating thin films on substrates in precise patterns, in which the substrate is placed in a high vacuum, and a guided beam of ionized molecules is fired at it, effectively allowing molecular-thin layers to be "painted" onto the substrate where required.

MOS - *see Metal-Oxide Semiconductor.*

Most-Significant Bit (MSB) - The binary digit, or bit, in a binary number that represents the most significant value (typically the left-hand bit).

Most-Significant Byte (MSB) - The byte in a multi-byte word that represents the most significant values (typically the left-hand byte).

Motherboard - A backplane containing active components such as integrated circuits.

MSB - *see Most-Significant Bit and Most-Significant Byte.*

MSI - *see Medium-Scale Integration.*

Multichip Module (MCM) - A generic name for a group of advanced interconnection

---

[10] Microwire is a registered trademark of Advanced Interconnection Technology (AIT), Islip, New York, USA.

and packaging technologies featuring unpackaged integrated circuits mounted directly onto a common substrate.

Multilayer - A printed circuit board constructed from a number of very thin single-sided and/or double-sided boards which are bonded together using a combination of temperature and pressure.

Multiplexer (digital) - A logic function that uses a binary value, or address, to select between a number of inputs and conveys the data from the selected input to the output.

Multiwire[11] - A trade name for the primary incarnation of discrete wire technology.

Nano - Unit qualifier (symbol = n) representing one thousandth of one millionth, or 10^{-9}. For example, 3nS stands for 3×10^{-9} seconds.

Nanobot - A molecular-sized robot (*see also Nano-Technology*).

Nanophase Materials - A form of matter which was only recently discovered, in which small clusters of atoms form the building blocks of a larger structure. These structures differ from those of naturally occurring crystals, in which individual atoms arrange themselves into a lattice.

Nano-Technology - Nano-technology is an elusive term that is used by different research-and-development teams to refer to whatever it is that they're working on at the time. However, regardless of their particular

area of interest, nano-technology always refers to something extremely small. One of the more outrageous branches of nano-technology that has been suggested as having potential in the future is that of micro-miniature electronic products that assemble themselves.

Negative-Edge - Beware, here be dragons. In this book a negative-edge is defined as a transition from a logic 1 to a logic 0. This definition is therefore consistent with the other definitions used throughout the book, namely those for active-high, active-low, assertion-level logic, positive-logic, and negative-logic.

It should be noted that some would define a negative-edge as a transition from a higher to a lower voltage level, passing through a threshold voltage level. However, this definition serves only to cause confusion when combined with negative-logic implementations.

Negative Ion - An atom or group of atoms with an extra electron.

Negative Logic - A convention which dictates the relationship between logical values and the physical voltages used to represent them. The more negative potential is considered to represent TRUE and the more positive potential is considered to represent FALSE.

Negative Resist - A process where ultraviolet radiation passing through the transparent areas of a mask causes the resist to be cured. The uncured areas are then removed using an appropriate solvent.

---

[11] Multiwire is a registered trademark of Advanced Interconnection Technology (AIT), Islip, New York, USA.

Negative-True - *see Negative Logic.*

Nibble - *see Nybble.*

NMOS (N-channel MOS) - Refers to the order in which the semiconductor is doped in a MOS device. That is, which structures are constructed as N-type versus P-type material.

Noble Gas - Gases whose outermost electron shells are completely filled with electrons. Such gases are extremely stable and it is difficult to coerce them to form compounds with other elements. There are six noble gasses: helium,[12] neon, argon, krypton, xenon, and radon. This group of elements were originally known as the *inert gasses*, but in the early 1960s it was found to be possible to combine krypton, xenon, and radon with fluorine to create compounds. Although helium, neon, and argon continue to resist, there is an increasing trend to refer to this group of gasses as *noble* rather than *inert*.

Noble Metal - Metals such as gold, silver, and platinum which are extremely inactive and are unaffected by air,[13] heat, moisture, and most solvents.

Noise - The miscellaneous rubbish that gets added to a signal on its journey through a circuit. Noise can be caused by capacitive or inductive coupling, or from externally generated interference.

Non-Volatile - A memory device which does not lose its data when power is removed from the system.

Non-Volatile RAM - A device which is generally formed from an SRAM die mounted in a package with a very small battery, or as a mixture of SRAM and EEPROM cells fabricated on the same die.

NPN (N-type - P-type - N-type) - Refers to the order in which the semiconductor is doped in a bipolar junction transistor.

N-type - A piece of semiconductor doped with impurities that make it amenable to donating electrons.

Nybble - A group of four binary digits, or bits.

Octal - Base-8 numbering system. Each octal digit can be directly mapped onto three binary digits, or bits.

Ohm - Unit of resistance. The Greek letter omega, Ω, is often used to represent ohms; for example, $1M\Omega$ indicates one million ohms.

One-Hot Encoding - A form of state assignment for state machines in which each state is represented by an individual state variable.

One-Time Programmable - A device such as a PAL, PLA, or PROM that can only

---

[12] The second most common element in the universe, after hydrogen.

[13] Before you start penning letters of protest, silver can be attacked by sulfur and sulfides which occur naturally in the atmosphere. Thus, the tarnishing seen on your mother's silver candlesticks is actually silver sulfide and not silver oxide.

be programmed a single time and whose contents cannot be subsequently erased.

Operating System - The collective name for the set of master programs that control the core operation and the base-level user-interface of a computer.

Optical Interconnect - The generic name for interconnection strategies based on opto-electronic systems, including fiber-optics, free-space, guided-wave, and holographic techniques.

Optical Lithography - A process in which radiation at optical wavelengths (usually in the ultraviolet range) is passed through a mask, and the resulting patterns are projected onto a layer of resist coating the substrate material.

Optical Mask - A sheet of material carrying patterns that are either transparent or opaque to the wavelengths used in an optical-lithographic process. Such a mask can carry hundreds of thousands of fine lines and geometric shapes.

Opto-Electronic - Refers to a system which combines optical and electronic components.

Organic Resist - A material which is used to coat a substrate and is then selectively cured to form an impervious layer. These materials are called organic because they are based on carbon compounds as are living creatures.

Organic Solvent - A solvent for organic materials such as those used to form organic resists.

Organic Substrate - Substrate materials such as FR4, in which woven glass fibers are bonded together with an epoxy. These materials are called organic because epoxies are based on carbon compounds as are living creatures.

Overglassing - One of the final stages in the integrated circuit fabrication process in which the entire surface of the wafer is coated with a layer of silicon dioxide or silicon nitride. This layer may also be referred to as the barrier layer or the passivation layer. An additional lithographic step is required to pattern holes in this layer to allow connections to be made to the pads.

Padcap - A special flavor of circuit board used for high-reliability military applications, also known as *pads-only-outer-layers*. Distinguished by the fact that the outer surfaces of the board have pads but no tracks. Signal layers are only created on the inner planes, and tracks are connected to the surface pads by vias.

Pad - An area of metallization on a substrate used for probing or to connect to a via, plated through-hole, or an external interconnect.

Pad Grid Array (PGA) - A packaging technology in which a device's external connections are arranged as an array of conducting pads on the base of the package.

Pad Stack - Refers to any pads, anti-pads, and thermal relief pads associated with a via or a plated through-hole as it passes through the layers forming the substrate.

Parallel-In Serial-Out (PISO) - Refers to a shift register in which the data is loaded in parallel and read out serially.

Parasitic Effects - The effects caused by undesired resistance, capacitance, or inductance inherent in the material or topology of a track or component.

Passivation Layer - *see Overglassing*.

Passive Trimming - A process in which a laser beam is used to trim components such as thick-film and thin-film resistors on an otherwise unpopulated and unpowered hybrid or multichip module substrate. Probes are placed at each end of a component to monitor its value while the laser evaporates some of the material forming the component.

Pass-Transistor Logic - A technique for connecting MOS transistors so that data signals pass between their source and drain terminals. Pass-transistor logic minimizes the number of transistors required to implement a function, and is typically employed by designers of cell libraries or full-custom integrated circuits.

PCB - *see Printed Circuit Board*.

Peta - Unit qualifier (symbol = P) representing one thousand million million, or 10^{15}. For example, 3PHz stands for 3×10^{15} Hertz.

PGA - *see Pin Grid Array and Pad Grid Array*.

PHB - *see Photochemical Hole Burning*.

Photochemical Hole Burning (PHB) - An optical memory technique, in which a laser in the visible waveband is directed at a microscopic point on the surface of a slice of glass which has been doped with organic dyes or rare-earth elements. The laser excites electrons in the glass such that they change the absorption characteristics of that area of the glass and leave a band, or hole, in the absorption spectrum.

Photo-Transistor - A special transistor which converts an optical input in the form of light into an equivalent electronic signal in the form of a voltage or current.

Pico - Unit qualifier (symbol = p) representing one millionth of one millionth, or 10^{-12}. For example, 3pS stands for 3×10^{-12} seconds.

Pin - *see Lead*.

Pin Grid Array (PGA) - A packaging technology in which a device's external connections are arranged as an array of conducting leads, or pins, on the base of the package.

PISO - *see Parallel-In Serial-Out*.

Place-Value - Refers to a numbering system in which the value of a particular digit depends both on the digit itself and its position in the number.

Plasma - A gaseous state in which the atoms or molecules are dissociated to form ions.

Plated Through-Hole - **(1)** A hole in a double-sided or multilayer board that is used to accommodate a through-hole component lead and is plated with copper. **(2)** An alternative name for the lead through-hole technique for populating circuit boards in which component leads are inserted into plated through-holes.

PMOS (P-channel MOS) - Refers to the order in which the semiconductor is doped in a MOS device. That is, which structures are constructed as P-type versus N-type material.

PNP (P-type - N-type - P-type) - Refers to the order in which the semiconductor is doped in a bipolar junction transistor.

Polysilicon Layer - An internal layer in an integrated circuit used to create the gate electrodes of MOS transistors. In addition to forming gate electrodes, the polysilicon layer can also be used to interconnect components. There may be several polysilicon layers separated by dielectric (insulating) layers.

Populating - The act of attaching components to a substrate.

Positive-Edge - Beware, here be dragons. In this book a positive-edge is defined as a transition from a logic 0 to a logic 1. This definition is therefore consistent with the other definitions used throughout the book, namely those for active-high, active-low, assertion-level logic, positive-logic, and negative-logic.

It should be noted that some would define a positive-edge as a transition from a lower to a higher voltage level, passing through a threshold voltage level. However, this definition serves only to cause confusion when combined with negative-logic implementations.

Positive Ion - An atom or group of atoms lacking an electron.

Positive Logic - A convention which dictates the relationship between logical values and the physical voltages used to represent them. The more positive potential is considered to represent TRUE and the more negative potential is considered to represent FALSE.

Positive Resist - A process where radiation passing through the transparent areas of a mask causes previously cured resist to be degraded. The degraded areas are then removed using an appropriate solvent.

Positive-True - *see Positive Logic*.

Power Plane - A conducting layer in or on the substrate providing power to the components. There may be several power planes separated by insulating layers.

Precedence of Operators - Determines the order in which operations are performed. For example, in standard arithmetic the multiplication operator has a higher precedence than the addition operator. Thus, in the equation $6 + 2 \times 3$, the multiplication is performed before the addition and the result is 12. Similarly, in Boolean Algebra, the AND operator has a higher precedence than the OR operator.

Prepreg - Non-conducting semi-cured layers of FR4 used to separate conducting layers in a multilayer circuit board.

Primitives - Simple logic functions such as BUF, NOT, AND, NAND, OR, NOR, XOR, and XNOR may be referred to as primitive logic gates or primitives.

Printed Circuit Board (PCB) - A type of circuit board which has conducting tracks superimposed, or "printed," on one or both sides, and may also contain internal signal layers and power and ground planes. An alternative name, Printed Wire Board (PWB), is commonly used in America.

Printed Wire Board (PWB) - A type of circuit board which has conducting tracks superimposed, or "printed," on one or both sides, and may also contain internal signal layers and power and ground planes. An alternative name, Printed Circuit Board (PCB), is predominantly used in Europe and Asia.

Product-of-Sums - A Boolean equation in which all the maxterms corresponding to the lines in the truth table for which the output is a logic 0 are combined using AND operators.

Product Term - A set of literals linked by an AND operator.

Programmable Array Logic (PAL)[14] - A programmable logic device in which the AND array is programmable but the OR array is predefined.

Programmable Logic Array (PLA) - The most user-configurable of the traditional programmable logic devices, because both the AND and OR arrays are programmable.

Programmable Logic Device (PLD) - The generic name for a device constructed in such a way that the designer can configure, or "program" it to perform a specific function.

Programmable Read-Only Memory (PROM) - A programmable logic device in which the OR array is programmable but the AND array is predefined. Usually considered to be a memory device whose contents can be electrically programmed (once) by the designer.

PROM - *see Programmable Read-Only Memory.*

Proposition - A statement that is either true or false with no ambiguity. For example, the proposition *"I just tipped a bucket of burning oil into your lap"* is either true or false, but there's no ambiguity about it.

Protein - A complex organic molecule formed from chains of amino acids, which are themselves formed from combinations of certain atoms, namely: carbon, hydrogen, nitrogen, oxygen, usually sulfur, and occasionally phosphorous or iron. Additionally, the chain of amino acids *"folds in on itself"* forming an extremely complex three-dimensional shape. Organic molecules have a number of useful properties, not the least that their structures are intrinsically *"self healing"* and reject contamination. Also, in addition to being extremely small, many organic molecules have excellent electronic properties. Unlike metallic conductors, they transfer energy by moving electron excitations from place to place rather than relocating entire electrons. This can result in switching speeds that are orders of magnitude faster than their semiconductor equivalents.

Protein Memory - A form of memory based on organic proteins *(see also Protein)*.

Protein Switch - A form of switch based on organic proteins *(see also Protein)*.

---

[14] PAL is a registered trademark of Monolithic Memories Inc.

Pseudo-Random - An artificial sequence of values that give the appearance of being random.

PTH - *see Plated Through-Hole.*

P-type - A piece of semiconductor doped with impurities that make it amenable to accepting electrons.

PWB - *see Printed Wire Board.*

QFP - *see Quad Flat Pack.*

Quad Flat Pack (QFP) - The most commonly used package in surface mount technology to achieve a high lead count in a small area. Leads are presented on all four sides of a thin square package.

Quantization - Part of the process by which an analog signal is converted into a series of digital values. First of all the analog signal is sampled at specific times. For each sample, the complete range of values that the analog signal can assume is divided into a set of discrete bands or quanta. Quantization refers to the process of determining which band the current sample falls into. *See also Sampling.*

Quinary - Base-5 numbering system.

Radix - Refers to the number of digits in a numbering system. For example, the decimal numbering system is said to be radix-10. May also be referred to as the "base."

RAM - *see Random Access Memory.*

Random-Access Memory (RAM) - A data-storage device from which data can be read out and new data can be written in. Unless otherwise indicated, the term RAM is typically taken to refer to a semiconductor device in the form of an integrated circuit.

Read-Only Memory (ROM) - A data-storage device from which data can be read out, but new data cannot be written in. Unless otherwise indicated, the term ROM is typically taken to refer to a semiconductor device in the form of an integrated circuit.

Read-Write Memory (RWM) - Alternative (and possibly more appropriate) name for a random access memory.

Real Estate - Refers to the amount of area available on a substrate.

Reconfigurable Hardware - A product whose function may be customized many times (*see also Configurable Hardware, Remotely Reconfigurable Hardware, Dynamically Reconfigurable Hardware, and Virtual Hardware*).

Reed-Müller Logic - Logic functions implemented using only XOR and XNOR gates.

Reflow Oven - An oven employing infrared radiation or hot air.

Reflow Soldering - A surface mount technology process in which the substrate and attached components are passed through a reflow oven to melt the solder paste.

Refractory Metal - Metals such as tungsten, titanium, and molybdenum which are capable of withstanding extremely high, or refractory, temperatures.

Remotely Reconfigurable Hardware - A product whose function may be customized remotely, by telephone or radio, while remaining resident in the system (*see also Configurable Hardware, Reconfigurable Hardware, Dynamically Reconfigurable Hardware, and Virtual Hardware*).

Resist - A material which is used to coat the substrate and is then selectively cured to form an impervious layer.

Resistor-Transistor Logic (RTL) - Logic gates implemented using particular configurations of resistors and bipolar junction transistors. For the majority of today's designers, resistor-transistor logic is of historical interest only.

Rigid Flex - Hybrid constructions which combine standard rigid circuit boards with flexible printed circuits, thereby reducing the component count, weight, and susceptibility to vibration of the circuit, and greatly increasing its reliability.

Rising-Edge - *see Positive-Edge.*

ROM - *see Read-Only Memory.*

RTL - *see Resistor-Transistor Logic.*

RWM - *see Read-Write Memory.*

Sampling - Part of the process by which an analog signal is converted into a series of digital values. Sampling refers to observing the value of the analog signal at specific times. *See also Quantization.*

Scalar Notation - A notation in which each signal is assigned a unique name; for example, $a3$, $a2$, $a1$, and $a0$. *See also Vector Notation.*

Scaling - A technique for making transistors switch faster by reducing their size. This strategy is known as scaling, because all of the transistors features are typically reduced by the same proportion.

Schematic - Common name for a circuit diagram.

Scrubbing - The process of vibrating two pieces of metal, or metal coated materials, at ultrasonic frequencies to create a friction weld.

Sea-of-Cells - Popular name for a channel-less gate array.

Sea-of-Gates. - Popular name for a channel-less gate array.

Seed Value - An initial value loaded into a linear feedback shift register or random number generator.

Semiconductor - A special class of material that can exhibit both conducting and insulating properties.

Sensor - A transducer that detects a physical quantity and converts it into a form suitable for processing. For example, a microphone is a sensor which detects sound and converts it into a corresponding voltage or current.

Sequential - A function whose output value depends not only on its current input values, but also on previous input values. That is, the output value depends on a sequence of input values.

Serial-In Parallel-Out (SIPO) - Refers to a shift register in which the data is loaded in serially and read out in parallel.

Serial-In Serial-Out (SISO) - Refers to a shift register in which the data is both loaded in and read out serially.

Sexagesimal - Base-60 numbering system.

Side-Emitting Laser Diode - A laser diode constricted at the edge of an integrated circuit's substrate such that, when power is applied, the resulting laser beam is emitted horizontally; that is, parallel to the surface of the substrate.

Signal Conditioning - Amplifying, filtering, or otherwise processing a signal.

Signal Layer - A layer carrying tracks in a circuit board, hybrid, or multichip module. *See also Wiring Layer*.

Signature - Refers to the checksum value from a cyclic-redundancy-check when used in the guided-probe form of functional test.

Signature Analysis - A guided-probe functional-test technique based on signatures.

Sign Bit - The most significant binary digit, or bit, of a signed binary number. If set to a logic 1, this bit represents a negative quantity.

Signed Binary Number - A binary number in which the most-significant bit is used to represent a negative quantity. Thus, a signed binary number can be used to represent both positive and negative values.

Sign-Magnitude - Negative numbers in standard arithmetic are typically represented in sign-magnitude form by prefixing the value with a minus sign; for example, −27.

For reasons of efficiency, computers rarely employ the sign-magnitude form. Instead, they use signed binary numbers to represent negative values.

Silicon Bumping - The process of depositing additional metallization on a die's pads to raise them fractionally above the level of the Barrier Layer.

Silicon Chip - Although a variety of semiconductor materials are available, the most commonly used is silicon and integrated circuits are popularly known as silicon chips, or simply chips.

Silicon Compiler - The program used in compiled cell technology to generate the masks used to create components and interconnections. May also be used to create datapath functions and memory functions.

Single-Sided - A printed circuit board with tracks on only one side.

Sintering - A process in which ultra-fine metal powders weld together at temperatures much lower than those required for larger pieces of the same materials.

SIPO - *see Serial-In Parallel-Out*.

SISO - *see Serial-In Serial-Out*.

Skin Effect - In the case of high frequency signals, electrons are only conducted on the outer surface, or skin, of a conductor. This phenomenon is known as the skin effect.

Small-Scale Integration (SSI) - Refers to the number of logic gates in a device. By one convention, small-scale integration represents a device containing 1 to 12 gates.

SMD - *see Surface Mount Device.*

SMT - *see Surface Mount Technology.*

SMOBC - *see Solder Mask Over Bare Copper.*

Soft Macro (Macro Function) - A logic function defined by the manufacturer of an application-specific integrated circuit. The function is described in terms of the simple functions provided in the cell library and connections between them. The assignment of cells to basic cells and the routing of the tracks is determined at the same time, and using the same tools, as for the other cells specified by the designer.

Software - Refers to intangible programs, or sequences of instructions, that are executed by hardware.

Solder - An alloy of tin and lead with a comparatively low melting point used to join less fusible metals. Typical solder contains 60% tin and 40% lead; increasing the proportion of lead results in a softer solder with a lower melting point, while decreasing the proportion of lead results in a harder solder with a higher melting point. Note that the solder used in a brazing process is of a different type, being a hard solder with a comparatively high melting point composed of an alloy of copper and zinc (brass).

Solder Bumping - A flipped chip technique in which spheres of solder are formed on the die's pads. The die is flipped and the solder bumps are brought into contact with corresponding pads on the substrate. When all the chips have been mounted on the substrate, the solder bumps are melted using reflow soldering or vapor-phase soldering.

Solder Mask - A layer applied to the surface of the substrate that prevents solder from sticking to any metallization except where holes are patterned into the mask.

Solder Mask Over Bare Copper (SMOBC) - A technique in which the solder mask is applied in advance of the tin-lead plating. This prevents solder from leaking under the mask when the tin-lead alloy melts during the process of attaching components to the board.

Space - Used to refer to the width of the gap between adjacent tracks.

SRAM - *see Static RAM.*

SSI - *see Small-Scale Integration.*

Standard Cell - An application-specific integrated circuit which, unlike a gate array, does not use the concept of a basic cell and does not have any prefabricated components. The manufacturer creates custom masks for every stage of the device's fabrication allowing each logic function to be created using the minimum number of transistors.

State Assignment - The process by which the states in a state machine are assigned to the binary patterns that are to be stored in the state variables.

State Diagram - A graphical representation of the operation of a state machine.

State Machine - The actual implementation (in hardware or software) of a function that can be considered to consist of a set of states through which it sequences.

Statement - A sentence that asserts or denies an attribute about an object or group of objects. For example, *"Your face resembles a cabbage."*

State Table - A tabular representation of the operation of a state machine. Similar to a truth table, but also includes the current state as an input and the next state as an output.

State Transition - An arc connecting two states in a state diagram.

State Variable - One of a set of registers whose values represent the current state occupied by a state machine.

Static Flex - A type of flexible printed circuit which can be manipulated into permanent three-dimensional shapes for applications such as calculators and high-tech cameras which require efficient use of volume and not just area.

Static RAM (SRAM) - A memory device in which each cell is formed from four or six transistors configured as a latch or a flip-flop. The term static is used because, once a value has been loaded into an SRAM cell, it will remain unchanged until it is explicitly altered or until power is removed from the device.

Steady State - A condition in which nothing is changing or happening.

Subatomic Erosion - *see Electromigration*.

Substrate - Generic name for the base layer of an integrated circuit, hybrid, multichip module, or circuit board. Substrates may be formed from a wide variety of materials, including semiconductors, ceramics, FR4

(fiberglass), glass, sapphire, or diamond depending on the application. Note that the term substrate has traditionally not been widely used in the circuit board world, at least not by the people who manufacture the boards. However, there is an increasing tendency to refer to a circuit board as a substrate by the people who populate the boards. The main reason for this is that circuit boards are often used as substrates in hybrids and multichip modules, and there is a trend towards a standard terminology across all forms of interconnection technology.

Subtractive Process - A process in which a substrate is first covered with conducting material, then any unwanted material is subsequently removed, or subtracted.

Sum-of-Products - A Boolean equation in which all the minterms corresponding to the lines in the truth table for which the output is a logic 1 are combined using OR operators.

Superconductor - A material with zero resistance to the flow of electric current.

Surface-Emitting Laser Diode - A laser diode constricted on an integrated circuit's substrate such that, when power is applied, the resulting laser beam is emitted directly away from the surface of the substrate.

Surface Mount Device (SMD) - A component whose packaging is designed for use with surface mount technology.

Surface Mount Technology (SMT) - A technique for populating hybrids, multichip modules, and circuit boards, in which packaged components are mounted directly onto the surface of the substrate. A layer of solder paste is screen printed onto the pads and the components are attached by pushing their leads into the paste. When all of the components have been attached, the solder paste is melted using either reflow soldering or vapor-phase soldering.

Symbolic Logic - A mathematical form in which propositions and their relationships may be represented symbolically using Boolean equations, truth tables, Karnaugh maps, or similar techniques.

Synchronous - **(1)** A signal whose data is not acknowledged or acted upon until the next active edge of a clock signal. **(2)** A system whose operation is synchronized by a clock signal.

TAB - *see Tape Automated Bonding.*

Tap- A register output which is used to generate the next data input to a linear feedback shift register.

Tape Automated Bonding (TAB) - A process in which transparent flexible tape has tracks created on its surface. The pads on unpackaged integrated circuits are attached to corresponding pads on the tape which is then stored in a reel. Silver-loaded epoxy is screen printed on the substrate at the site where the device is to be located and onto the pads to which the device's leads are to be connected. The reel of TAB tape is fed through an automatic machine which pushes the device and the TAB leads into the epoxy. When the silver-loaded epoxy is cured using reflow soldering or vapor-phase soldering, it forms electrical connections between the TAB leads and the pads on the substrate.

Tera - Unit qualifier (symbol = T) representing one million million, or 10^{12}. For example, 3THz stands for 3×10^{12} Hertz.

Tertiary - Base-3 numbering system.

Tertiary Digit - A numeral in the tertiary scale of notation. A tertiary digit can adopt one of three states: 0, 1, or 2. Often abbreviated to "trit."

Tertiary Logic - An experimental technology in which logic gates are based on three distinct voltage levels. The three voltages are used to represent the tertiary digits 0, 1, and 2, and their logical equivalents FALSE, TRUE, and MAYBE.

Thermal Relief Pad - A special pattern etched around a via or a plated through-hole to connect it into a power or ground plane. A thermal relief pad is necessary to prevent too much heat being absorbed into the power or ground plane when the board is being soldered.

Thermal Tracking - Typically used to refer to the problems associated with optical interconnection systems whose alignment may be disturbed by changes in temperature.

Thick-Film Process - A process used in the manufacture of hybrids and multichip modules in which signal and dielectric (insulating) layers are screen-printed onto the substrate.

Thin-Film Process - A process used in the manufacture of hybrids and multichip modules in which signal layers and dielectric (insulating) layers are created using opto-lithographic techniques.

Through-Hole - *see Lead Through-Hole, Plated Through-Hole, and Through-Hole Via.*

Through-Hole Via - A via that passes all the way through the substrate.

Thru-Hole - A commonly used abbreviation of "through-hole."

Time-Of-Flight - The time taken for a signal to propagate from one logic gate or opto-electronic component to another.

Tin-Lead Plating - An electroless plating process in which exposed areas of copper on a circuit board are coated with a layer of tin-lead alloy. The alloy is used to prevent the copper from oxidizing and provides protection against contamination.

Tinning - *see Tin-Lead Plating.*

Toggle - Refers to the contents or outputs of a logic function switching to the inverse of their previous logic values.

Trace - *see Track.*

Track - A conducting connection between electronic components. May also be called a trace or a signal. In the case of integrated circuits, such interconnections are often referred to collectively as metallization.

Transducer - A device that converts input energy of one form into output energy of another.

Transistor - A three-terminal semiconductor device that, in the digital world, can be considered to operate like a switch.

Tri-State Function - A function whose output can adopt three states: 0, 1, and Z

(high-impedance). The function does not drive any value in the Z state and, in many respects, may be considered to be disconnected from the rest of the circuit.

Trit - Abbreviation of tertiary digit. A tertiary digit can adopt one of three values: 0, 1, or 2.

Truth Table - A convenient way to represent the operation of a digital circuit as columns of input values and their corresponding output responses.

TTL - *see Transistor-Transistor Logic.*

Transistor-Transistor Logic (TTL) - Logic gates implemented using particular configurations of bipolar junction transistors.

ULA - *see Uncommitted Logic Array.*

Ultra-Large-Scale Integration (ULSI) - Refers to the number of logic gates in a device. By one convention, ultra-large-scale integration represents a device containing a million or more gates.

ULSI - *see Ultra-Large-Scale Integration.*

Uncommitted Logic Array (ULA) - One of the original names used to refer to gate array devices. This term has largely fallen into disuse.

Unsigned Binary Number - A binary number in which all the bits are used to represent positive quantities. Thus, an unsigned binary number can only be used to represent positive values.

Vapor-Phase Soldering - A surface mount process in which a substrate carrying components attached by solder paste is lowered into the vapor-cloud of a tank containing boiling hydrocarbons. This melts the solder paste thereby forming good electrical connections. However, vapor-phase soldering is becoming increasingly less popular due to environmental concerns.

Vaporware - Refers to either hardware or software that exist only in the minds of the people who are trying to sell them to you.

Vector Notation - A notation in which a single name is used to reference a group of signals, and individual signals within the group are referenced by means of an index; for example, a[3], a[2], a[1], and a[0].

Note that this type of vector notation is in no way related to the algebraic concept of vector notation for Cartesian 2-space or 3-space. Some people prefer to use the phrase binary group notation, but the phrase vector notation is commonly used by practicing electronic engineers. Additionally, this concept of a vector is commonly used in the context of electronics tools such as schematic capture packages, logic simulators, and graphical waveform displays.

Very-Large-Scale Integration (VLSI) - Refers to the number of logic gates in a device. By one convention, very-large-scale integration represents a device containing 1,000 to 999,999 gates.

Via - A hole filled or lined with a conducting material which is used to link two or more conducting layers in a substrate.

Virtual Hardware - An extension of dynamically configurable hardware based on a new generation of FPGAs which were

introduced around the beginning of 1994. In addition to supporting the dynamic reconfiguration of selected portions of the internal logic, these devices also feature: no disruption to the device's inputs and outputs; no disruption to the system-level clocking; the continued operation of any portions of the device that are not undergoing reconfiguration; and no disruption to the contents of internal registers during reconfiguration, even in the area being reconfigured *(see also Configurable Hardware, Reconfigurable Hardware, Remotely Reconfigurable Hardware, and Dynamically Reconfigurable Hardware)*.

Virtual Memory - A trick used by a computer's operating system to pretend that it has access to more memory than is actually available. For example, a program running on the computer may require ten mega-bytes to store its data, but the computer may have only five mega-bytes of memory available. To get around this problem, whenever the program attempts to access a memory location that does not physically exist, the operating system performs a slight-of-hand and exchanges some of the contents in the memory with data on the hard disk.

Virus - *see Computer Virus*.

VLSI - *see Very-Large-Scale Integration*.

Volatile - Refers to a memory device which loses any data it contains when power is removed from the system; for example, random-access memory in the form of SRAM or DRAM.

Wafer - A paper-thin slice cut from a cylindrical crystal of pure semiconductor.

Wafer Probing - The process of testing individual integrated circuits while they still form part of a wafer. An automated tester places probes on the device's pads, applies power to the power pads, injects a series of signals into the input pads, and monitors the corresponding signals returned from the output pads.

Waveguide - A transparent path bounded by non-transparent, reflective areas, which is fabricated directly onto the surface of a substrate. Used in the optical interconnection strategy known as guided-wave.

Wave Soldering - A process used to solder circuit boards populated with through-hole components. A machine creates a wave[15] of hot, liquid solder which travels across the surface of the tank. The populated circuit boards are passed over the wave soldering machine on a conveyer belt. The velocity of the conveyer belt is carefully controlled and synchronized such that the solder wave brushes across the bottom of the board only once.

Wire Bonding - The process of connecting the pads on an unpackaged integrated circuit to corresponding pads on a substrate using wires that are finer than a human hair. Wire bonding may also be used to connect the pads on an unpackaged integrated circuit, hybrid, or multi-chip module to the leads of the component package.

Wiring Layer - A layer carrying wires in a discrete wired board. *See also Signal Layer.*

Word - A group of signals or logic functions performing a common task and carrying or storing similar data; for example, a value on the data bus could be referred to as a data word.

X-Ray Lithography - Similar in principle to optical lithography, but capable of constructing much finer features due to the shorter wavelengths involved. However, X-ray lithography requires an intense source of X-rays, is more difficult to use, and is considerably more expensive than optical lithography.

Yield - The number of devices that work as planned, specified as a percentage of the total number actually fabricated.

---

[15] A large ripple, actually.

Dramatis Personae

‡ = footnote

Index

---

bold number = key definition, [‡] = footnote

---

bold number = key definition,   [‡] = footnote

bold number = key definition, [‡] = footnote

(continued)

bold number = key definition, ‡ = footnote

---

[1] Cache Logic is a trademark of Atmel Corporation, San Jose, CA, USA

bold number = key definition, [‡] = footnote

bold number = key definition, ‡ = footnote

bold number = key definition, [‡] = footnote

E

---

bold number = key definition, ‡ = footnote

bold number = key definition, ‡ = footnote

---

[2] FPIC is a trademark of Aptix Corporation.

  bold number = key definition, [‡] = footnote

bold number = key definition, [‡] = footnote

---

bold number = key definition, [‡] = footnote

---

bold number = key definition, ‡ = footnote

bold number = key definition, [‡] = footnote

---

bold number = key definition, [‡] = footnote

---

bold number = key definition, [‡] = footnote

---

bold number = key definition, [‡] = footnote

P

---

[3] PAL is a registered trademark of Monolithic Memories Inc.

bold number = key definition, [‡] = footnote

---

bold number = key definition, ‡ = footnote

bold number = key definition, ‡ = footnote

---

bold number = key definition, [‡] = footnote

---

bold number = key definition, [‡] = footnote

---

bold number = key definition, ‡ = footnote

---

bold number = key definition, [‡] = footnote

bold number = key definition, ‡ = footnote